Manned
Spaceflight
Log

Tim Furniss

Manned
Spaceflight
Log

New edition

JANE'S

First published in the United Kingdom in 1983
This edition published in 1986 by
Jane's Publishing Company Limited
238 City Road, London EC1V 2PU

ISBN 0 7106 0402 5

Distributed in the Philippines and the USA and its dependencies by
Jane's Publishing Inc,
115 5th Avenue,
New York, NY 10003

Computer typesetting by Method Limited,
Epping, Essex

Printed in the United Kingdom by
Biddles Limited, Guildford, Surrey

Acknowledgements

I would like to express my thanks to a number of people who have helped with the preparation of this second edition of *Manned Spaceflight Log*: my wife Sue for her support, suggestions and comments; Rex Hall and Neville Kidger, both expert Soviet space analysts, who provided me with updated flight times and other invaluable information; Dave Shayler of Astro Information Service, High Farm Road, Halesowen, West Midlands B62 9RX, England, for his help in providing data and photographs; Lisa Vazquez of Nasa's Johnson Space Centre, Ed Harrison of the Kennedy Space Centre, Mark Jones of Novosti and Ludmila Pakhomova of Tass for providing photographs; and finally my friend and editor Brendan Gallagher for his usual enthusiastic encouragement and professional work.

If you are interested in keeping abreast of manned spaceflight developments, why not join the British Interplanetary Society (27/29, South Lambeth Road, London SW8 1SZ) and the National Space Society (West Wing Suite 203, 600 Maryland Avenue SW, Washington DC 20024)? Both of these organisations publish magazines designed to keep professionals and enthusiasts in touch with what is happening in space.

Tim Furniss
Epsom, Surrey
August 1986

To my son, Tom

About the author

Tim Furniss, pictured here at the Yuri Gagarin Cosmonaut Training Centre with its commander, Lt-Gen Georgi Beregovoi, writes for *Flight International* and contributes to *Space, Satellite and Space Technology*, *Space World*, *Space Flight News* and the national press. He is also associate editor of *The Space Report*, a fortnightly business bulletin, and operates a photo library. The author of *Space Shuttle Log* and nearly 20 other books, he is a regular broadcaster on TV and radio programmes such as *Newsnight*, *Saturday Superstore* and *Today*. He is a member of the British Space Society, a Fellow of the British Interplanetary Society and a member of the US National Space Society. He has covered Apollo and Shuttle launches from the Kennedy Space Centre, visited the Yuri Gagarin cosmonaut training centre in Star City as a guest of Soviet space chiefs, and interviewed many astronauts, cosmonauts and leading industrialists and administrators. Married, with one son, he lives in Epsom, Surrey, where he was born in 1948. His hobbies include running, squash and wicketkeeping for Tadworth Cricket Club.

Introduction

Manned Spaceflight Log has been updated and extensively reillustrated to cover the flights that have taken place during the 25 years between Vostok 1 and as close as possible to the present. The result is, I hope, a readable book which also presents the facts in a readily usable manner. I have also tried to express the individual "personality" of each mission.

Manned Spaceflight Log includes the 13 missions by the X-15 rocket aircraft that exceeded an altitude of 50 miles between 1962 and 1968. It is thus the first book to recognise the achievements of the X-15 in reaching space. Although the Fédération Aéronautique Internationale (FAI) regards the altitude of 100km (62 miles) as the beginning of space, the *Log* includes the X-15 flights that exceeded 50 miles because the USAF awarded astronaut wings to those of its pilots who flew some of these missions. The X-15 flights are therefore listed as "astro-flights," along with all the other missions in this book. Then each rocket-boosted flight into FAI-recognised space is listed as a "spaceflight". Finally, because some of these latter flights were sub-orbital, a third classification, "Earth orbit," is included. Moon-flight data are treated in the same way. The flight times of the Space Shuttle missions run from launch to main-gear touchdown, not nose-gear touchdown or wheels stop.

Name: Vostok 1
Sequence: 1st astro-flight, 1st spaceflight, 1st Earth orbit
Launch date: April 12, 1961
Launch site: Tyuratam, USSR
Launch vehicle: A1 (SL-3)
Flight type: Earth orbit
Flight time: 1hr 48min
Spacecraft weight: 10,419lb
Crew: Lt Yuri Alexeyevich Gagarin, 27, Soviet Air Force

At the Tyuratam Cosmodrome, about 230 miles south-east of Baikonur in the Soviet Republic of Kazakhstan, on April 12, 1961, an ordinary-looking tourist coach stopped beneath a skeletal steel tower. Within its metallic embrace stood a rocket about 125ft tall. Two men dressed in orange flight suits and white pressure helmets got out of the coach, shook hands and briefly clicked their helmets together. One of the men climbed a flight of steps towards a lift and turned before entering it. He held both hands high and bade farewell to his back-up and the watching technicians.

When the lift reached the top of the rocket the man got out and walked towards a small circular hatchway. He squeezed himself through, feet first, and lay on the couch within. He was in a cramped spherical capsule, surrounded by dials and switches. The door was closed behind him. In an hour the man's name was to be on everyone's lips.

Yuri Alexeyevich Gagarin was born on March 9, 1934, in the village of Klushino, near Smolensk. His father was a carpenter. In 1955 Gagarin joined a flight training school, flew solo and made parachute jumps. He then entered the Air Force and learned to fly jets at Orenburg on the River Ural before becoming one of the first cosmonaut trainees in 1960.

Models of his Vostok spacecraft had been successfully tested three times out of five attempts under the guise of Sputniks 4, 5, 6, 9 and 10. The 7ft 5in-diameter capsule was attached by four metal straps to an instrument section and a retro-rocket package. The complete craft was about 14ft long and weighed about 10,000lb in orbit, while the capsule alone weighed just 1,765lb.

Vostok's A1 launch vehicle was basically the SS-6 Sapwood ballistic missile with a Luna second stage, so called because it was first used to launch the Luna 1 Moon probe. This vehicle, and the spacecraft itself, were not revealed to the world until 1967. Compared with American vehicles, A1 looked distinctly odd. It consisted of a hammer-headed central stage with four RD 108 engines and four vernier motors, plus four tapered strap-on stages, each with four RD 107 engines and two verniers. The strap-ons were ejected during the ascent, so that A1 was in effect a 2½-stage booster. Total thrust amounted to more than one million pounds. After a short delay during which a faulty valve was replaced, the A1 ascended into space at 9.07 am Moscow time. The film of the lift-off, first shown in the West seven years later, is memorable, with the rocket's shadow seen flickering across the flat steppe country around Tyuratam. "Off we go!" shouted the jubilant Gagarin.

Breathing a cabin atmosphere of oxygen and nitrogen at sea-level pressure, Gagarin entered orbit a few minutes later. "The sky", he reported, "looks very, very dark and the Earth is bluish". Weightlessness was very relaxing, he said, and made a welcome change to the g forces of acceleration during

Yuri Gagarin (Novosti)

A replica of Vostok 1. Gagarin was housed in the spherical flight and re-entry capsule. Beneath this is the retro-rocket pack (Novosti)

7

launch. Vostok reached a maximum altitude of 203 miles during its orbit over South America, the South Atlantic and Africa at an inclination of 65° to the equator. This orbit would have decayed naturally within ten days, the operational lifetime of the spacecraft. But a retro-rocket was automatically fired by radio signal when Vostok was making its first pass over Africa, and re-entry began.

The capsule separated from the retro-package and plunged into the atmosphere with its heavier, base, side first, an all-over ablative heatshield protecting its precious cargo. This period of high temperatures and risks lasted for a few minutes, and then at 13,000ft a braking drogue parachute was deployed, followed by the main chute at 8,000ft. Before this Gagarin had ejected and landed separately from the capsule, which hit the ground in a field at Smelovaka, near Saratov, watched by a cow and two bemused farm workers.

Gagarin's one history-making orbit had lasted 108min and he had travelled about 25,400 miles. His flight had been announced by the Russians while he was still in orbit. The newspapers of the world went wild, hailing Gagarin as the Columbus of the space age.

Gagarin, who was promoted to Major during the flight, did not make any more voyages of discovery. After many years as a fine ambassador for the Soviet space effort, and on the verge of a second career as a cosmonaut, he was killed in a air crash on March 27, 1968. The Soviet hierarchy paid him the supreme posthumous tribute, ordering that his ashes be placed in the Kremlin Wall on Moscow's Red Square.

The historic moment of ignition. After full thrust had been reached, the restraining arms sprang back and the rocket was free to rise. The spherical Vostok 1 capsule was protected by a conical payload shroud which was ejected during the launch (Novosti)

Yuri Gagarin shortly before his death in 1968 (Novosti)

Freedom 7/Mercury-Redstone 3   May 5, 1961   Flight 2

Name: *Freedom 7*/Mercury-Redstone 3/Mercury craft 7
Sequence: 2nd astro-flight, 2nd spaceflight
Launch date: May 5, 1961
Launch site: Pad 5, Cape Canaveral, USA
Launch vehicle: Redstone (MR-7 launcher)
Flight type: Sub-orbital
Flight time: 15min 28sec
Spacecraft weight: 2,845lb
Crew: Cdr Alan Bartlett Shepard Jr, 38, USN

The United States was not to know it at the time, but the Mercury spacecraft was rather more sophisticated than its Soviet counterpart. However, when America attempted to fly a man in Mercury for the first time the objective was not even a single orbit. Indeed, Nasa had yet to put even an unmanned Mercury into orbit as preparations for Mercury-Redstone 3 went ahead in the shadow of Vostok 1.

Mercury was a bell-shaped capsule about 9ft 5in high and with a maximum diameter across the ablative heatshield base of 6ft 1in. At lift-off, with a launch-escape system (LES) rocket at its apex, it weighed about 4,200lb, falling to 2,900lb once the LES had been jettisoned. The attitude of the capsule in space could be changed by the release of short bursts of hydrogen peroxide gas from 18 thrusters located on the craft. These movements could be controlled by the Automatic Stability and Control System (ASCS), which acted as the craft's "autopilot"; from the ground through the Rate Stabilisation and Control System (RSCS); or by the astronaut himself through a hand controller connected to a fly-by-wire system.

The first manned flight was to be one of three sub-orbital, "up-and-down" missions to man-rate Mercury for orbital flight. The launch vehicle was to be the 80ft-high Redstone intermediate-range ballistic missile (IRBM), with a thrust of 78,000lb.

Three astronauts from the group of seven chosen in 1959 had been selected to make these flights, which had been reduced in number to three from the planned seven. The original intention had been that each man would gain some experience before one of the seven attempted orbital flights. Now it was intended to tackle orbital flight after just three sub-orbital "lobs". The three pioneer astronauts were Cdr Alan Shepard, Capt Virgil Grissom and Lt Col John Glenn. The identity of American spaceman No 1 was kept secret at first, only to be inadvertently revealed after a flight postponement on May 2, three days before the eventual launch date. It was Shepard.

Cdr Shepard was born in East Derry, New Hampshire, on November 18, 1923, and graduated with a science degree from the US Naval Academy in 1944. Following destroyer service in the war he gained pilot's wings in 1947 and later became a test pilot, working on in-flight refuelling techniques. In 1958 he was appointed an air readiness officer for the Commander-in-Chief Atlantic Fleet and his name was automatically put forward, among those of hundreds of other military test pilots, as a potential astronaut. The "Mercury Seven", as the successful candidates became known, were introduced to the world in April 1959, and Shepard was among them.

Above A Redstone missile soars into the sky over Pad 5 at Cape Canaveral carrying the Mercury capsule Freedom 7. ***Below** America's first spaceman, Alan Shepard, retrieves checklists from his capsule aboard the recovery ship USS* Lake Champlain *after his history-making flight (Nasa)*

On May 5, 1961, Shepard lay inside the capsule he had named *Freedom 7* for fully 4hr 14min before lift-off. Unlike Vostok's it took place in full view of the world, covered live by television and radio. Mercury-Redstone 3 was finally launched at 9.34 am Cape Canaveral time. "I have lift-off, the clock is started," was the first of Shepard's 78 laconic voice communications during the 15min flight. With the aid of a one-second burn by three posigrade thrusters on the retro-pack, the capsule separated from the Redstone about five minutes later and soared into a trajectory that took it to a height of 116 miles and a maximum speed of 5,180mph. During Shepard's brief 4min 45sec of weightlessness he test-fired the sets of attitude thrusters for 40sec, causing movements in yaw, pitch and roll and thus becoming the first man to control his craft in space. His only glimpse of the Earth came not through his small porthole but in black and white by means of a periscope. "What a beautiful view," he said, almost as if trying to convince himself.

Although the retro-rockets were not needed to get *Freedom 7* back to Earth, they were test-fired to prove them for orbital flight. Correctly oriented so that its heatshield bore the brunt of the searing current of air molecules, *Freedom 7* safely negotiated the hazards of re-entry. Sailors on the recovery vessel, the USS *Lake Champlain*, saw Shepard's descent and splashdown beneath a single parachute into the Atlantic Ocean 297 miles downrange from the Cape. Just before landing the heatshield was dropped 4ft, pulling out a rubberised fibreglass landing bag designed to reduce shock. "Boy, what a ride!" said Shepard.

Almost as soon as he had been hauled aboard the recovery helicopter, preparations were going ahead at Cape Canaveral for Mercury-Redstone 4, a repeat performance of Shepard's flight. And on May 25, 1961, not long after pinning a Distinguished Service Medal onto Shepard's lounge suit, President Kennedy sent America to the Moon by launching Project Apollo. Alan Shepard would go there too, though only after a long campaign to overcome a physical problem and regain astronaut status.

Liberty Bell 7/Mercury-Redstone 4   July 21, 1961   Flight 3

Name: *Liberty Bell 7*/Mercury-Redstone 4/Mercury craft 11
Sequence: 3rd astro-flight, 3rd spaceflight
Launch date: July 21, 1961
Launch site: Pad 5, Cape Canaveral, USA
Launch vehicle: Redstone (MR-8 launcher)
Flight type: Sub-orbital
Flight time: 15min 37sec
Spacecraft weight: 2,836lb
Crew: Capt Virgil Ivan "Gus" Grissom, 35, USAF

Virgil Grissom's first words after his spaceflight were hardly history-making: "Give me something to blow my nose," he growled, "my head is full of seawater." Grissom had just lost his spaceship.

Born in Mitchell, Indiana, on April 23, 1926, Grissom received a degree in mechanical engineering in 1950. He then entered the Air Force, won his wings in March 1951, and subsequently flew 100 combat missions in F-86 Sabres during the Korean War. Gaining a degree in aeronautical engineering at the Air Force Institute of Technology in 1955, he later became a test pilot and joined the Mercury team in 1959.

Grissom called his spacecraft *Liberty Bell 7* after the famous Liberty Bell, joking that a crack ought to be painted on the side of the capsule. Mercury-Redstone 4 rose into an almost cloudless sky at 7.20 am on July 21, 1961, and was soon arcing its way over the Atlantic Ocean. *Liberty Bell 7*

Now you see it, now you don't. **Below** Liberty Bell 7 *is shipping water fast after losing its hatch.* **Below right** *The recovery helicopter's undercarriage touches the water as the pilot struggles vainly to lift the water-filled capsule. Grissom, meanwhile, was almost drowning* (Nasa)

Gus Grissom seen before entering Liberty Bell 7. (Nasa via Astro Information Service)

reached a maximum speed of 5,310mph and a height of 118.5 miles. During the flight Grissom varied the attitude of the craft and fired the retro-rockets. He also enjoyed the view from a new rectangular window.

The spacecraft splashed down in the Atlantic 303 miles downrange of Cape Canaveral at T+15min 37sec. As the recovery helicopters hovered over the capsule, Grissom completed his final checks and prepared the hatch for opening by arming the explosive bolts. Suddenly the hatch blew off, the sea rushed in and Grissom jumped out. Both the capsule and the astronaut began to sink, the latter because he had not sealed the hose connection of his suit. While the crew of the helicopter attached a line to the capsule, the floundering Grissom was becoming quite distraught as another machine hovered overhead to allow one crewman to take photographs of him. Finally a line was lowered and the unfortunate astronaut was hauled 30ft underwater before emerging from the ocean. Meanwhile, the other helicopter seemed to be losing a tug-of-war with the spacecraft, which was shipping a lot of water and getting heavier by the second. A warning light in the cockpit indicated to the helicopter pilot that he should let go of the capsule, which immediately sank in thousands of feet of water. Later it was discovered that the warning light was a false alarm and that *Liberty Bell 7* could have been rescued.

The unhappy Grissom was taken aboard the prime recovery ship, the *USS Randolph*, still completely mystified by the mishap. To the day he died he swore he was not responsible for the problem: "I was just lying there and it blew".

Vostok 2   August 6, 1961   Flight 4

Name: Vostok 2
Sequence: 4th astro-flight, 4th spaceflight, 2nd Earth orbit
Launch date: August 6, 1961
Launch site: Tyuratam, USSR
Launch vehicle: A1 (SL-3)
Flight type: Earth orbit
Flight time: 1 day 1hr 18min
Spacecraft weight: 10,431lb
Crew: Capt Gherman Stepanovich Titov, 25, Soviet Air Force

In August 1961, as the United States was still struggling to put an unmanned Mercury capsule into orbit, preparations were afoot in the Soviet Union for a Vostok flight that would last over a day. The massive leap from one to 17 orbits resulted from the need to land the craft in the prime recovery zone, which would be overflown every 24hr or 17 orbits.

The pilot of Vostok 2 was Gherman Titov, aged 25 years 329 days on the day of the flight. Over 20 years later this remains the youngest at which anyone has gone into space. He was also the first man to sleep in space – and the first to be

Gherman Titov, who made his flight at the age of 25, the youngest that anyone has ever gone into space (Novosti)

Vostok spacecraft during final assembly (Novosti)

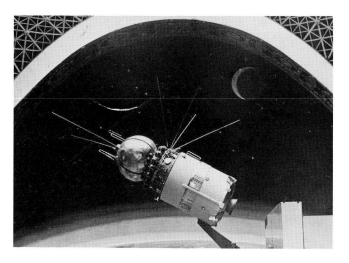

Vostok attached to the final stage of the launcher (Novosti)

spacesick. Titov was born in a log cabin in the village of Verkhneye Zhilino in the Altai region of the Soviet Union on September 11, 1935. When he was 18 he learned to fly at Kustenai, and from 1955-1957 he was a cadet at the Stalingrad Air Force Pilots School, from which he graduated with distinction. Titov joined the cosmonaut team in 1960. A wrist injury sustained when he was a youth remained undetected for many years, and after his Vostok mission it was conceded by the Soviet space authorities that if they had known about it they would not have let him make the flight.

Almost as soon as he reached his 111/160-mile orbit, at about 9.40 am Moscow time, Titov reported feeling "magnificent" and set to work manoeuvring the spacecraft. Later it was revealed that he felt anything but magnificent. Weight-

lessness had affected his inner-ear balance mechanism, causing him to feel sick. He remained so for much of the flight, and experienced further difficulty when his arms floated disconcertingly as he tried to sleep on the seventh orbit. He solved this problem by tucking his arms under his seat belt.

Re-entry was uneventful and, as planned, Titov ejected at 22,000ft from the capsule, separating from the seat at 8,000ft and landing by parachute. Vostok 2 alighted near the Vostok 1 landing site.

Titov had travelled 436,656 miles in one day 1hr 18min. At the time of writing Titov had not made another flight, unlike many of his colleagues.

Friendship 7/Mercury-Atlas 6   February 20, 1962   Flight 5

Name: *Friendship 7*/Mercury-Atlas 6/Mercury craft 13
Sequence: 5th astro-flight, 5th spaceflight, 3rd Earth orbit
Launch date: February 20, 1962
Launch site: Pad 14, Cape Canaveral, USA
Launch vehicle: Atlas 109D
Flight type: Earth orbit
Flight time: 4hr 55min 23sec
Spacecraft weight: 2,987lb
Crew: Lt-Col John Herschel Glenn Jr, 40, USMC

As chimpanzee Enos was' being recovered from the Atlantic Ocean on November 29, 1961, after completing a trouble-filled second orbital test flight of the Mercury capsule, Nasa confirmed that the third manned Redstone ballistic mission would be scrubbed and its pilot reassigned to the first manned orbital flight. So John Glenn, instead of making what would have been the most anonymous Mercury mission, flew the one that captured worldwide attention and made him an

American hero. It was also the flight that the seven original astronauts coveted most.

The flight was scheduled to take place on December 20, 1961, but was postponed first to January 16, 1962, then January 23 and January 27. On the last occasion Glenn lay in the tiny capsule for nearly six hours as the countdown was held again and again. His torture was ended by bad weather and when he finally emerged from the capsule he said quietly: "Oh well, there'll be another day."

Glenn was born in Cambridge, Ohio, on July 18, 1921, and entered the US Navy as a cadet in 1942. After receiving his pilot's wings he entered the Marine Corps and flew 59 combat missions over the Pacific during the Second World War. During the Korean conflict Glenn flew 63 combat missions, downing three MIG-15s and earning an Air Medal and eighteen clusters. In July 1957 he set a transcontinental record in Project Bullet by flying an F-8 Crusader fighter from Los Angeles to New York in 3hr 23min. Glenn was back-up to both Shepard and Grissom, and he called his Mercury capsule *Friendship 7*.

Friendship 7 was rescheduled once more, for February 4 and finally for February 20. Then at 9.47 am the gleaming 95ft Mercury-Atlas ignited on Cape Canaveral's Pad 14 and Glenn headed for orbit in a worldwide glare of publicity. The astronaut had been up since 2.20 am and in the capsule since 6.03 am. Sitting atop an initial thrust of 367,000lb from the Atlas's three engines, and undergoing only 7g compared with the 11g imposed by the Redstone, *Friendship 7* soared to a height of 97.6 miles. The capsule then separated from the rocket and turned around under automatic control.

"The view is tremendous, it is beautiful … a beautiful sight," said Glenn as he looked back towards Florida while heading backwards to Africa. Unlike Titov, Glenn was happy with weightlessness. "I have nothing but a fine feeling," he said. "It feels very normal and very good." The astronaut controlled *Friendship 7* manually, took photographs, reported instrument readings and saw strange "fireflies" floating outside the spacecraft. "What did they say, John?," asked ground control in a jocular allusion to the current round of UFO fever. But then came problems when the automatic controls of the craft malfunctioned. A sticking yaw thruster meant that Glenn continually had to return to manual control to correct the drift.

Meanwhile, another and potentially more serious problem was developing. Unknown to Glenn, technicians monitoring the flight had recorded a signal indicating that the heatshield on *Friendship 7* was loose. If the heatshield slipped off during re-entry Glenn would be burned to ashes in seconds. His only chance seemed to be to keep the jettisonable retro-pack on during re-entry in the hope that it would secure the heatshield if indeed it was loose. None of this was relayed to Glenn, and he was puzzled by the request to leave the pack on after retrofire.

Over the Pacific, at a point half-way between Hawaii and Los Angeles and and approaching the end of the third orbit, the retros were fired, giving Glenn the sensation that he was "going clear back to Hawaii". As he came through the upper layers of the atmosphere he was reminded to leave the retro-pack on. At that point his heart rate rose to 132 beats per minute. Soon he saw debris hurtling past the capsule. Was the heatshield loose? The transmissions from *Friendship 7* ceased, as predicted, when the craft became enveloped in incandescent gases raised to a temperature of 1,650°C as a result of aerodynamic heating at 25 miles altitude and 15,000mph. The wait was unbearable, but then Glenn's voice broke through the transmission crackle: "This is *Friendship 7*. A real fireball outside. My condition is good but that was a real fireball – boy! I had chunks of that retro-pack breaking off all the way through".

Glenn splashed down at T+4hr 55min 23sec, 40 miles uprange from the prime recovery ship, the USS *Randolph*, stationed 210 miles north-west of San Juan, Puerto Rico. He and his capsule were hauled aboard the USS *Noa*. Glenn hurt his hand when blowing the capsule's exit hatch but was otherwise completely unhurt. He emerged at 3.23 pm and later witnessed his fourth sunset on a day in which he had travelled 75,679 miles. America went wild over its new hero and Glenn was given one of the biggest New York tickertape welcomes ever.

Above right *Breakfast for John Glenn before he puts on his spacesuit and prepares to become America's first man in orbit* (Nasa via Astro Information Service)

Right *Atlas 109D leaves Pad 14 at Cape Canaveral, taking with it Mercury capsule* Friendship 7 *and astronaut John Glenn* (Nasa)

Name: *Aurora 7*/Mercury-Atlas 7/Mercury craft 18
Sequence: 6th astro-flight, 6th spaceflight, 4th Earth orbit
Launch date: May 24, 1962
Launch site: Pad 14, Cape Canaveral, USA
Launch vehicle: Atlas 107D
Flight type: Earth orbit
Flight time: 4hr 56min 5sec
Spacecraft weight: 2,975lb
Crew: Lt-Cdr Malcolm Scott Carpenter, 37, USN

Scott Carpenter had served as John Glenn's back-up, so when on March 17, 1962, the prime pilot for the Mercury-Atlas 7 flight, Deke Slayton, lost his flight status because of a heart flutter, it was Carpenter and not Slayton's back-up, Wally Schirra, who took the mission. Carpenter had been trained for a three-orbit mission, and that's what MA-7 was to be. It was cruel luck for Slayton, who, had the original Mercury-Redstone sub-orbital schedule been adhered to, would have become the first American in orbit.

Born in Boulder, Colorado, on May 1, 1925, Carpenter studied aeronautical engineering at Colorado State University, graduating in 1949. He then joined the Navy, earning his wings in April 1951. Carpenter was assigned to anti-submarine patrol duties during the Korean War and was an intelligence officer on the aircraft carrier USS *Hornet* when he was posted to Project Mercury.

Mercury-Atlas 7 was to be a scientific mission. Mercury had been tested in orbit and Carpenter's job was to make exterior observations of the Earth's weather, land masses, horizon, sunrises and sunsets. On May 24, 1962, he awoke at 1.15 am and by 4.40 am was lying in the cabin of the capsule he had called *Aurora 7* after his hometown street. The launch had been held for 45min before the Atlas finally ignited and rose through sea mist at 7.45 am towards a 100/167-mile orbit. "Great Scott!" read one banner newspaper headline.

"I'm weightless!" Carpenter exclaimed as the capsule separated from its rocket and turned around. He then fully tested the manual control systems and ejected a tethered balloon from the nose of the spacecraft. The balloon, designed to provide the astronaut with a target for experiments covering the perception of distances and the visibility of objects in space, unfortunately failed to inflate correctly.

Carpenter admitted after the flight that he had found the view, particularly the sunrises and sunsets, so awe-inspiring that he fell behind schedule and became careless in the rush to catch up. He was behind time as he finished the first of the three planned orbits and his attention was diverted further by a spacesuit temperature-control problem that caused him to sweat profusely, and a control malfunction that called for repeated manual corrections. These extra firings depleted the attitude-control fuel supply, and then the harassed astronaut made his big mistake. While switching from automatic to manual in preparation for the important manoeuvre preceding retrofire, Carpenter forgot to turn the first system off, resulting in a critical waste of fuel.

As he neared the end of the flight Carpenter was taking photographs of a sunrise, talking to ground control about re-entry procedures, storing equipment and discovering the source of the "fireflies" seen by Glenn. Accidentally hitting

Malcolm Scott Carpenter, the sixth man in space and the fourth to orbit the Earth (Nasa via Astro Information Service)

the roof of *Aurora 7*, Carpenter saw small ice particles scattering from the side of the craft. They were formed, he concluded, from water venting from the spacecraft. More fuel was wasted as he manoeuvred the craft to take a better look. A yaw error and lack of fuel led Carpenter to align *Aurora 7* at the wrong angle for re-entry, and he then fired the retro-rockets five seconds late. The cumulative effect of these errors was an eventful re-entry during which Carpenter encountered severe oscillations and was extremely concerned for his safety, and an overshoot of 250 miles.

Aurora 7 splashed down at T+4hr 56min, miles from the prime recovery ship, USS *Intrepid*, and 125 miles north-east of Puerto Rico. Newspaper and other reports indicated that the astronaut was lost for two hours and that nobody knew whether he had even survived the re-entry. Evidence suggests however that he was spotted 36min after splashdown but not actually picked up for another hour and a half. Whatever the case, the "lost astronaut" story diverted attention from what in Nasa's view had not been a good flight.

Carpenter was picked up from his liferaft alongside *Aurora 7* by USS *Pierce*. He was obviously elated by his 76,021-mile flight, but Nasa did not seem to share that opinion and Carpenter soon left the programme. If the truth be known, Carpenter was more a victim of politics within the space corps than of his own frailty.

Carpenter pictured after his pre-flight medical on launch day (Nasa via Astro Information Service)

Atlas 107D lifts off into the early-morning mist at the start of the flight of Aurora 7 (Nasa)

X-15 No 3, Flight No 7   July 17, 1962   Flight 7

Name: X-15 No 3, Flight No 7
Sequence: 7th astro-flight
Flight date: July 17, 1962
Take-off site: Edwards Air Force Base, California, USA
Launch vehicle: B-52 carrier aircraft
Flight type: Sub-orbital
Flight time: 10min 20sec
Spacecraft weight: About 34,000lb
Crew: Maj Robert Michael White, 38, USAF

The Space Shuttle was not in fact the first reusable manned spacecraft. That honour went to the X-15 rocket aircraft, which flew into sub-orbital space a number of times. The X-15 was a forerunner of the Space Shuttle and an extremely valuable research tool.

Work on the X-15 began in 1954 following the dramatic high-speed, high-altitude flights of earlier X-series research aircraft such as the X-1 and X-2. Measuring 50ft long with a wingspan of 22ft, the X-15 had a wedge-shaped vertical tail and thin, stubby wings. It weighed about 14,000lb empty and

34,000lb fully fuelled. The manually controlled XLR rocket engine had a thrust of about 60,000lb. At the hypersonic speeds attained by the X-15 aerodynamic heating was an acute structural problem. This was solved by applying a heat-resistant skin made of Inconel-X nickel-steel alloy over a titanium and stainless-steel structure.

The X-15, of which three examples were built, was designed to provide data on the aerodynamics, structural and control problems, and physiological aspects of high-speed, high-altitude flight. It was also later to act as a testbed, carrying various scientific experiments beyond the Earth's atmosphere. For the flight in the lower atmosphere the X-15 embodied conventional aerodynamic controls. In space the pilot used eight hydrogen peroxide thrusters on the nose to control pitch and yaw, and four on the wings for roll control.

The X-15 was air-launched from beneath the wing of a B-52 at about 45,000ft and 500mph. The rocket motor ignited after the drop and fired for about 80sec. Two flight profiles were used: one for high altitude, with a steep rate of climb, the other for high speed in a level altitude. The remainder of the

10-11min flight was powerless, ending with a 200mph landing. The X-15 was first flown in 1959, and of its first 44 flights no fewer than 13 would have failed had a pilot not been aboard. The 62nd flight in the programme, and the seventh by the No 3 aircraft, was destined to go into the history books.

Robert White flew this mission, his 15th on the X-15. He was born in New York City on July 6, 1924, and gained degrees in electrical engineering and business administration. White was one of the few spacemen to have fought in World War II. Flying P-51s, he was shot down on his 52nd mission, captured by the Germans and released at the war's end. After a brief spell out of the Air Force, by 1954 White had graduated from the Experimental Test Pilot School at Edwards Air Force Base. In 1957 White was selected as the back-up USAF X-15 pilot and became the prime pilot when the legendary Iven Kincheloe was killed in an air crash in July 1958. He flew his first X-15 mission on April 13, 1960.

On July 17, 1962, after taking off from Edwards Air Force Base, B-52 project pilot Maj John Allavie and RAF Gp Capt Harry Archer took Bob White, strapped into the X-15 cockpit, to an altitude of 45,000ft. Reaching a point over Smith Ranch, Nevada, the mother ship turned west until it was over Delamar. Then the X-15 dropped away from the B-52 and its rocket motor ignited. White's mission was to attain an altitude of 280,000ft and then to try out the reaction control system. The rocket aircraft went into a 41° climb and its engine fired for one second longer than planned, pushing the maximum speed to 3,832mph (Mach 5.45), 284mph faster than planned. The X-15 soared to a height of 314,750ft (59.61 miles), briefly becoming a spacecraft and making its pilot the 7th man above 50 miles. Because the USAF had stipulated that any pilot exceeding this height would receive astronaut wings, White became the first of 13 "unofficial" X-15 spacemen. The flight was ratified by the FAI as a world altitude record for an aircraft, but was still three miles short of the 62-mile (100km) limit beyond which lay FAI-defined space.

At the zenith of his trajectory White could see a sweep of territory from San Francisco to Mexico. He also saw something else. "Through my left windshield," he said later, "I saw something that looked like a piece of paper, the size of my hand, tumbling slowly outside the plane. It was greyish in colour and about 30 to 40ft away." It is thought that this was a piece of ice from the control thrusters, a large "Glenn firefly" in fact.

After three minutes of zero g White used the reaction control system to set up the X-15 for re-entry, during which he was subjected to a sustained 5g and his aircraft to ultra-high temperatures. Nearing Edwards at 18,000ft, White used the aerodynamic controls to initiate a corkscrew dive. He pulled out at 1,000ft, jettisoned the lower tail of the aircraft, lowered the landing gear and came in for a landing on the dry bed of Rogers Lake at 200mph. His epic mission had lasted 11min.

Robert White, the fifth American to fly above 50 miles (USAF)

The X-15's engine lights up and Robert White heads for near-space. This was taken at the start of White's flight on November 9, 1961, when he reached a speed of 4,093 mph (Nasa)

Vostok 3   August 11, 1962   Flight 8
Vostok 4   August 12, 1962   Flight 9

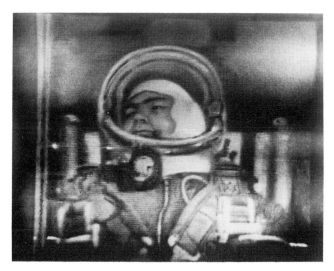

Andrian Nikolayev, shown inside Vostok 3 during the first ever telecast from space (Novosti)

*A Vostok rocket arriving at the pad **above** and poised for launch **below*** (Novosti)

Name: Vostok 3
Sequence: 8th astro-flight, 7th spaceflight, 5th Earth orbit
Launch date: August 11, 1962
Launch site: Tyuratam, USSR
Launch vehicle: A1 (SL-3)
Flight type: Earth orbit
Flight time: 3 days 22hr 22min
Spacecraft weight: 10,410lb
Crew: Maj Andrian Grigoryevich Nikolayev, 32, Soviet Air Force

Name: Vostok 4
Sequence: 9th astro-flight, 8th spaceflight, 6th Earth orbit
Launch date: August 12, 1962
Launch site: Tyuratam, USSR
Launch vehicle: A1 (SL-3)
Flight type: Earth orbit
Flight time: 2 days 22hr 57min
Spacecraft weight: 10,423lb
Crew: Lt-Col Pavel Romanovich Popovich, 31 Soviet Air Force

The Soviet Union's standing in space was high in 1962, and it soared even higher after the next spectacular, which was a brilliant technical and propaganda success. But its substance was less impressive than its appearance. Because the West did not know much about the Russian space programme, the dual flight of Vostoks 3 and 4 was given a degree of press coverage that it did not in fact warrant.

The first bachelor spaceman kissed his girlfriend goodbye at the base of the Vostok 3 launch pad on August 11, 1962, and was on his way into space at 11.30 am. Andrian Nikolayev was born in Shorshaly in the Volga basin, about 700 miles east of Moscow, on September 5, 1929. He studied to become a physician, worked as a lumberjack, entered the Air Force as a radio operator, trained as a gunner and won his pilot's wings in 1954. He joined the cosmonaut team in 1960 and was Gherman Titov's back-up.

The quiet Nikolayev entered a 112/145-mile orbit and settled down to what was assumed to be an extended flight lasting about four or five days. As Vostok 3, code-named *Falcon*, passed over Tyuratam at 11.02 am a day later on its 16th orbit, Vostok 4 *Golden Eagle* climbed up towards it, carrying Pavel Popovich. Born in the village of Uzin near Kiev on October 5, 1930, Popovich worked as a shepherd and studied building construction before entering the Air Force. Before becoming a cosmonaut in 1960 he won the Order of the Red Star for flying a secret mission over the Arctic.

Vostok 4 entered its 112/158-mile orbit within minutes, and for a short time the two manned spacecraft were four miles apart, with Nikolayev able to see his colleague's craft. Although the Vostoks had no ability to manoeuvre from orbit to orbit and effect a true rendezvous, the "meeting" was hailed by the Western press as the "first rendezvous on the road to the Moon". "Nik and Pop meet in space" was one headline. Another newspaper even gave space to an eminent British scientist when he predicted a Russian lunar landing

Pavel Popovich demonstrates weightlessness inside Vostok 4 (Novosti)

by 1965. But all this ignored the fact that the ability to rendezvous and dock, so vital to a Moon flight, had yet to be demonstrated.

The first ever television broadcast from space enabled viewers to see Nikolayev in his cabin, which was large enough for him to undo his straps and float freely. Medical experiments and other observations were conducted by both cosmonauts, and they even tried proper food instead of the "toothpaste tube" products eaten on earlier flights. Nikolayev and Popovich sampled cutlets, pies, juices and fruit with great success.

During the flight the spacecraft drifted further apart, so that by Vostok 3's 64th orbit they were separated by 1,700 miles. Both cosmonauts returned to Earth on August 15, landing about 120 miles apart and to the south of the town of Karaganda. Vostok 4 clocked up 48 orbits and 1,230,230 miles in 70hr 57min, while Nikolayev in Vostok 3 achieved a record 64 orbits and 1,639,190 miles in 94hr 22min.

Sigma 7/Mercury-Atlas 8   October 3, 1962   Flight 10

Name: *Sigma 7*/Mercury-Atlas 8/Mercury craft 16
Sequence: 10th astro-flight, 9th spaceflight, 7th Earth orbit
Launch date: October 3, 1962
Launch site: Pad 14, Cape Canaveral, USA
Launch vehicle: Atlas 113D
Flight type: Earth orbit
Flight time: 9hr 13min 11sec
Spacecraft weight: 3,029lb
Crew: Cdr Walter Marty Schirra Jr, 39, USN

Wally Schirra's first trip into space, aboard Mercury capsule *Sigma 7*, has been described as a textbook mission. Compared with Carpenter's eventful flight it certainly was straightforward, even though the astronaut had to cope with a couple of problems before everything settled down. Perhaps the only major surprise was the mission's duration, which at 5¾ orbits hardly represented a significant advance on *Aurora 7*.

Up from the launch pad, down to the sea. **Right** *Sigma 7 is launched from Pad 14 at Cape Canaveral and,* **below,** *is recovered from the Pacific Ocean (Nasa)*

Schirra was born in Hackensack, New Jersey, on March 12, 1923, and graduated from high school in 1940 before entering engineering college. He joined the US Navy Academy a year later and graduated in 1945, seeing brief service aboard USS *Alaska*. He learned to fly at Pensacola, Florida, and joined the 71st Navy Fighter Squadron. During the Korean War Schirra served in a USAF squadron, flying 90 combat missions, downing two MiG-15s and earning the DFC and two Air Medals. He then became a test pilot.

Mercury-Atlas 8 began at 8.15 am on October 3, 1962. In Britain television viewers saw the launch three hours later via the Telstar communications satellite. The Atlas D vehicle developed an alarming roll rate during the launch and an abort was almost ordered. Once in the 100/176-mile orbit, the highest by a US astronaut to that date, and travelling at 17,558mph, Schirra ran into difficulties with his spacesuit, which was overheating because of a faulty valve. This nearly

Below Wally Schirra in space. *Below right* Schirra suiting up. *"I've got nothing else to do today," he said. "I might as well take a trip into space."* (Nasa via Astro Information Service)

led to an early return after just one orbit, but fortunately Schirra was the Mercury spacesuit expert and he was able to solve the problem. A 30in balloon was deployed on a 100ft tether and this time the experiment, first tried by Carpenter, was a success. But Schirra's flight did not capture the headlines. Indeed, one policeman at Cape Canaveral was heard to describe it as "dullsville". The Apollo project would run into a similar waning of public interest a decade later.

Once established in orbit, Schirra turned off all the attitude-control systems and allowed *Sigma 7* to drift in a slow roll for 99min. After 8hr 56min 22sec of zero g the retros fired and *Sigma 7* descended towards the Midway Islands for a Pacific splashdown. Schirra experienced a maximum re-entry deceleration of 8.1g and splashed down in the Pacific – the first astronaut to do so – just over four miles from the USS *Kearsarge* and 295 miles north-east of Midway Island. The craft and the astronaut were hauled onto the deck and Schirra emerged fit and smiling. Of his 9hr 13min, 153,900-mile flight Schirra said: "It was a routine, textbook flight. I was able to accomplish everything I wanted during the flight. It was a honey of a machine."

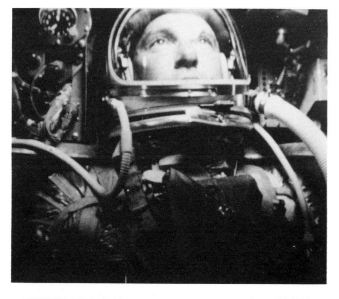

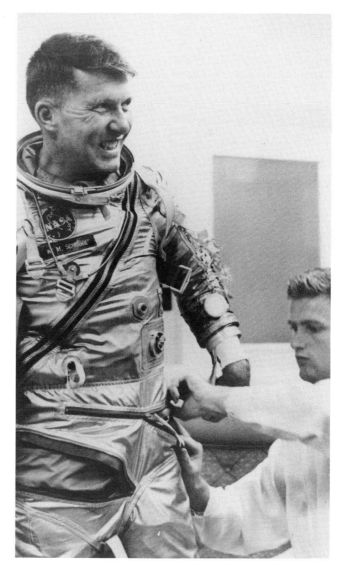

X-15 No 3, Flight No 14   January 17, 1963   Flight 11

Name: X-15 No 3, Flight No 14
Sequence: 11th astro-flight
Flight date: January 17, 1963
Take-off site: Edwards Air Force Base, California, USA
Launch vehicle: B-52 carrier aircraft
Flight type: Sub-orbital
Flight time: 9min 23sec
Spacecraft weight: About 34,000lb
Crew: Joseph Albert Walker, 41

The next spaceman was a civilian test pilot who had already become a legend in aerospace history. Born in Washington DC on February 20, 1921, Joe Walker gained a physics degree in 1942 and joined the USAF, flying P-38s during the Second World War. He left the USAF at the end of the war with a DFC and an Air Medal with seven clusters. Walker joined NACA in March 1945 and served as project pilot on such high-performance research types as the Skyrocket, X-1, X-3, X-4 and X-5. In 1959 he was named chief X-15 pilot for the newly created Nasa, and he flew the first US Government X-15 mission on March 25, 1960. By June 1962 he had reached a record speed of 4,100 mph in the rocket aircraft.

On January 17, 1963, Walker flew his 18th X-15 mission, the 14th by X-15 No 3, and the 77th in the programme. His brief was to study the handling of the aircraft without its ventral fin and at extreme altitudes, and to conduct an infra-red experiment. The B-52 carrier aircraft took the X-15 to height over Delamar Lake, Nevada, before Walker dropped clear, lit the engine and screamed away to 3,677mph at 146,000ft (Mach 5.47). The engine cut out at 81sec and Walker reached an apogee of 271,700ft (51 miles).

As a civilian Joe Walker didn't receive the USAF astronaut wings, but he qualified nonetheless for the title of "unofficial astronaut".

Joe Walker, the 11th man above 50 miles, stands next to an X-15 (Nasa)

Faith 7/Mercury-Atlas 9   May 15, 1963   Flight 12

Name: *Faith 7*/Mercury-Atlas 9/Mercury craft 20
Sequence: 12th astro-flight, 10th spaceflight, 8th Earth orbit
Launch date: May 15, 1963
Launch site: Pad 14, Cape Canaveral, USA
Launch vehicle: Atlas 130D
Flight type: Earth orbit
Flight time: 1 day 10hr 19min 49sec
Spacecraft weight: 3,033lb
Crew: Maj Leroy Gordon Cooper Jr, 36, USAF

Lying in Mercury capsule *Faith 7* on May 15, 1963, astronaut Gordon Cooper seemed the most relaxed of all the launch team. He even fell asleep during the countdown.

Cooper, destined to become the last Mercury spaceman as a result of the grounding of Deke Slayton, was also at 36 the youngest of the astronauts. He was born in Shawnee, Oklahoma, on March 6, 1927, and had the unusual distinction of being connected with all four military services. A member of the Marines in 1945, he attended the US Naval Academy and was a member of the Presidential Guard. He then went to the University of Hawaii before receiving a commission to join the Army. Finally he joined the Air Force, learning to fly in 1949 and serving for a time in Germany. After graduating with a science degree in aeronautical engineering from the Air Force Institute of Technology in 1956 he became a test pilot and later an instructor at Edwards Air Force Base, California.

Water in the diesel engine of the launch tower caused a postponement on May 14, preventing technicians from moving the structure away from the rocket. But there were no hitches the following day, when at 9.04 am Cooper lifted off for a 100/166-mile orbit. The first in-flight television from an American capsule was rather disappointing. British viewers who saw the pictures the same day via the Relay comsat could only just make out the face of the astronaut and unrecognisable vistas of the globe when Cooper pointed the camera out of the window. What wasn't disappointing, especially to military men, was Cooper's amazing ability to see comparatively small features on Earth, including smoke from a log cabin in the Himalayas, the wake of a ship and a aircraft contrail. Corroborating Cooper's claims would be a wealth of high-quality photographs taken with a wide range of cameras.

Cooper encountered and overcame the perennial space-suit temperature problem, and unsuccessfully deployed a tethered balloon. He had more success with a flashing beacon ejected from the nose of the capsule to test his ability to sight objects in space, an important prelude to rendezvous. Then, with his hands safely tucked into the seat straps, Cooper slept soundly for 7½hr. He remained remarkably relaxed throughout the flight, and at one point he was using so little oxygen that capcom Wally Schirra joked: "You can stop holding your breath".

Then trouble hit *Faith 7*. First, on the 19th orbit, Cooper noticed a light which indicated that the craft was no longer weightless, suggesting that it was making a premature re-entry. In tracing the cause of the problem it became clear that there had been a major malfunction of the automatic control system. This meant that Cooper would have to perform an entirely manual re-entry. "The task is difficult, but the man is trained," a Nasa spokesman explained. The subsequent perfect splashdown of *Faith 7* just 7,000 yards from USS *Kearsarge* in the Pacific, 80 miles south-east of Midway, after 21¾ orbits seemed to emphasise the importance of human control of spacecraft. Apart from slight giddiness, Cooper seemed in perfect health as he was helped from his capsule after a journey of some 583,469 miles that·had put manned spaceflight back into the limelight once more.

After some talk of Alan Shepard, Cooper's back-up, flying a seventh Mercury on a 72hr mission the project officially ended on June 12, 1963, after totalling two days 5hr 55min 27sec of manned spaceflight at a cost of $392 million.

The Mercury astronauts used to say that they didn't get into the spacecraft, they put it on. Here Gordon Cooper illustrates the problem (Nasa)

This side view of the Atlas launching of Faith 7 *shows clearly the vernier motors firing to stabilise the rocket (Nasa)*

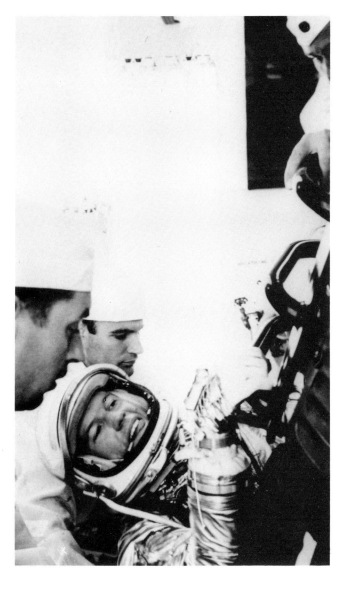

Vostok 5
Vostok 6

June 14, 1963
June 16, 1963

Flight 13
Flight 14

Name: Vostok 5
Sequence: 13th astro-flight, 11th spaceflight, 9th Earth orbit
Launch date: June 14, 1963
Launch site: Tyuratam, USSR
Launch vehicle: A1 (SL-3)
Flight type: Earth orbit
Flight time: 4 days 23hr 6min
Spacecraft weight: 10,441lb
Crew: Lt-Col Valeri Fyodorovich Bykovsky, 28, Soviet Air Force

Name: Vostok 6
Sequence: 14th astro-flight, 12th spaceflight, 10th Earth orbit
Launch date: June 16, 1963
Launch site: Tyuratam, USSR
Launch vehicle: A1 (SL-3)
Flight type: Earth orbit
Flight time: 2 days 22hr 50min
Spacecraft weight: 10,392lb
Crew: 2nd Lt Valentina Vladimirovna Tereshkova, 26, Soviet Air Force

During the early days of June 1963 rumours emanating from Moscow spoke of the impending launch of a woman into space. Day by day, more details of the flight plan emerged: a man would be launched first, to be joined in space by a woman the following day.

The man was Valeri Bykovsky, whose launch aboard Vostok 5 had been delayed from June 13 by bad weather. The next attempt, at 2.59 pm the following day, was a success, though it was scarcely heeded by a world eagerly awaiting the spacewoman who was to join him on June 15, on his 17th orbit. That day came and went without a second launch, giving rise to speculation that the intended woman cosmonaut was incapacitated. If that was the case, her place was taken by Valentina Tereshkova, who on the next pass of Vostok 5 over Tyuratam lifted off aboard Vostok 6 at 12.30 pm on June 16 to join Bykovsky. The newspapers went to town on the headlines on June 17 as the "Adam and Eve of space," as Bykovsky and Tereshkova were dubbed, came to within three miles of each other during Vostok 6's first orbit.

Bykovsky, code-named *Hawk* for the purposes of the joint flight, was born in the Moscow suburb of Pavlov-Posad on August 2, 1934, and learned to fly when he was 17. After being named as a cosmonaut in 1960 he was the first to try out the centrifuge and other devices used to train Soviet spacemen. His first major assignment was as the Vostok 3 back-up pilot.

Tereshkova was born on March 6, 1937, in Maslennikovo, Yaroslavl. A cotton mill worker and amateur parachutist with 126 jumps to her credit, Tereshkova joined the cosmonaut corps in March 1962.

Valeri Bykovsky, still the solo spaceflight record-holder (Novosti)

Bykovsky preparing for a parachute jump in training (Novosti)

Valentina Tereshkova, the first woman in space and the 10th human being in orbit (Novosti)

Tereshkova about to board Vostok 6 (Novosti)

The charred and battered Vostok 6 capsule after landing. Tereshkova, seated on the right, is watched by curious farmworkers (Novosti)

The fact that no other woman followed Tereshkova into orbit until Svetlana Savitskaya in Soyuz T-6 in 1982 lends weight to the assertion that her flight was simply one more stunt in Premier Khruschev's space spectacular campaign. But she certainly earned the world's admiration, particularly that of the Communist Women's League, which conveniently was about to start its annual conference in Moscow. Reports from unofficial sources indicated that Tereshkova was not happy in space. She is said to have pleaded to come home but was kept in orbit for three days, returning after 48 orbits and about 1,242,800 miles. Tereshkova's official flight time was 70hr 50min and she landed 390 miles north-east of Karaganda.

Bykovsky, the forgotten man of the joint mission, landed shortly afterwards 337 miles north-west of Karaganda on the same day, having clocked up a flight time of four days 23hr

6min, 81 orbits and two million miles. Twenty years later Bykovsky still holds the record for the longest solo spaceflight, with no likelihood of his losing that honour in these days of multi-man missions.

Like Mercury, the Vostok programme ended after the sixth flight, bringing Soviet manned spaceflight time to 15 days 22hr 21min, 13 days ahead of the US total. The Americans had a long way to catch up, and the West was wondering what new feat the Russians had up their sleeves.

There was a happier footnote to Valentina Tereshkova's apparently miserable mission. The identity of the woman who said goodbye to Nikolayev at the launch pad in 1962 was revealed in 1964 when he and Tereshkova married in Moscow. A year later Tereshkova gave birth to their daughter Yelena, the world's first "space baby".

X-15 No 3, Flight No 20   June 27, 1963   Flight 15

Name: X-15 No 3, Flight No 20
Sequence: 15th astro-flight
Flight date: June 27, 1963
Take-off site: Edwards Air Force Base, California, USA
Launch vehicle: B-52 carrier aircraft
Flight type: Sub-orbital
Flight time: 10min 28sec
Spacecraft weight: About 34,000lb
Crew: Maj Robert A. Rushworth, 39, USAF

June 27, 1963, was the date for the 87th X-15 mission, the 20th flight of the No 3 aircraft, and the 14th X-15 flight by the pilot, Maj Robert Rushworth. It also made Rushworth the 15th man above 50 miles and earned him USAF astronaut wings.

Born in Madison, Maine, on October 9, 1924, Rushworth studied mechanical engineering before joining the USAF. He flew transport missions over China during the Second World War and combat missions in Korea. After the Korean War

Robert Rushworth, the 15th man above 50 miles (USAF)

Rushworth graduated from the USAF Institute of Technology and in 1959 was selected as an X-15 pilot. He flew his first mission on November 4, 1960.

His 14th flight was designed to acquaint him with high-altitude handling and phenomena. The X-15 was released over Delamar, Nevada, and the engine fired at 100 per cent for 80sec, taking the aircraft to a top speed of 3,425mph (Mach 4.89) and an altitude of 285,000ft (almost 55 miles). Mission duration was 10.46min.

An X-15 rocket aircraft in flight (Nasa via Astro Information Service)

X-15 No 3, Flight No 21   July 19, 1963   Flight 16
X-15 No 3, Flight No 22   August 22, 1963   Flight 17

Name: X-15 No 3, Flights No 21 and 22
Sequence: 16th and 17th astro-flights
Flight date: July 19, 1963; August 22, 1963
Take-off site: Edwards Air Force Base, California, USA
Launch vehicle: B-52 carrier aircraft
Flight type: Sub-orbital
Flight times: 11min 24sec, 11min 8sec
Spacecraft weight: About 34,000lb
Crew: Joseph Albert Walker, 42

Joe Walker, the first man above 50 miles twice and three times

Joe Walker unofficially became the first man to make two spaceflights and the first to make three, but he never got the astronaut's wings to prove it.

Walker made his 24th X-15 flight on July 19, 1963. It was the 90th X-15 mission and the 21st by X-15 No 3. The aircraft was to be flown to high altitude for studies of the expansion of the airframe during re-entry with the ventral fin removed. Walker was also to deploy and tow a nitrogen-filled balloon, and to conduct horizon-scanning, photometer, infra-red and ultra-violet observations.

He was launched over Smith Ranch, Nevada, and the engine fired at full thrust for 85sec, two seconds longer than planned, taking the X-15 to a height of 347,800ft (65.3 miles). "I thought my altimeter had gone whacky," he said after a flight that took him to a maximum speed of 3,710mph (Mach 5.50) and lasted 11.4min. According to the FAI definition, space begins at an altitude of 62 miles (100km). Walker could

therefore have claimed official astronaut status. He could have repeated this claim on August 22, the date of the next X-15 flight.

The objective of the 91st flight in the programme was an altitude of 360,000ft. Walker was launched over Smith Ranch and achieved a maximum speed of 3,794mph (Mach 5.58) before the engine shut down at T+85.5sec. The rocket

aircraft, sporting the legend "Little Joe the Second" on its nose, reached an altitude of 354,200ft (66.75 miles) during the 11.1min flight, the highest the X-15 was to achieve.

Walker left the programme on this high note and in 1965 tested the Apollo lunar landing research vehicle. Soon afterwards he died in particularly tragic circumstances. Flying an F-104 in formation with an XB-70 Valkyrie during a public relations demonstration on June 8, 1966, he collided with the tail of the supersonic bomber when, it is believed, he was momentarily distracted. The F-104 disintegrated and Walker was killed. The XB-70 also crashed, killing one of its pilots.

An X-15 coming in to land. The lower ventral fin has been jettisoned and skids deployed for touchdown on the dry lake bed at Edwards Air Force Base, California (Nasa)

Voskhod 1   October 12, 1964   Flight 18

Name: Voskhod 1
Sequence: 18th astro-flight, 13th spaceflight, 11th Earth orbit
Launch date: October 12, 1964
Launch site: Tyuratam, USSR
Launch vehicle: A2 (SL-4)
Flight type: Earth orbit
Flight time: 1 day 0hr 17min 3sec
Spacecraft weight: 11,731lb
Crew: Col Vladimir Mikhailovich Komarov, 37, Soviet Air Force, commander
Konstantin Petrovich Feoktistov, 38, scientist
Lt Boris Borisovich Yegorov, 27, Soviet Air Force, doctor

On October 13, 1964, the daily newspapers in the West carried banner headlines and features about the mammoth new three-man spaceship that starred in Russia's latest space spectacular. No two-man crew had yet been into space, and here were the Russians launching three. Spectacular indeed – or was it?

The truth of the matter is that the unfortunate cosmonauts were crammed together into a stripped Vostok one-man

spacecraft, with neither spacesuits nor ejection seats. It was probably the most perilous mission undertaken, and all in the cause of Premier Khruschev's quest for new spectaculars to maintain the Soviet Union's psychological lead over the United States in the Cold War. America was planning to launch a two-man Gemini in the near future, so Khruschev wanted his scientists to launch not two but three cosmonauts.

The composition of the crew gave another clue to the propaganda nature of the flight. There was the required pilot, of course, and then two crewmen who had only begun cosmonaut training six months earlier, having been plucked from the support team. Thus it was that the Soviet Union could claim that the Voskhod "spaceship" counted a scientist and a doctor amongst its crew.

Pilot-cosmonaut Vladimir Komarov was born in Moscow on March 16, 1927, and entered pilot training when he was 15. He later graduated from four Air Force colleges and joined the first group of cosmonauts in 1960. Komarov was Popovich's back-up on Vostok 4 before being temporarily grounded by a minor heart complaint similar to that which affected America's Deke Slayton. Unlike Slayton, though, the

Left to right *Boris Yegorov, Konstantin Feoktistov and Vladimir Komarov after emerging from the Voskhod 1 capsule* (Novosti)

Rare view of a Voskhod launch (Novosti)

Russian made it back to flight status in good time.

Konstantin Feoktistov, the scientist, was the man whose job it was to make Vostok a relatively safe vehicle for three men. Born on February 26, 1926, in Voronezh, he was wounded during the war and subsequently studied engineering in Moscow.

Doctor Boris Yegorov was at 27 the youngest of the team. Born in Moscow on November 26, 1937, he was the son of a prominent brain surgeon. He studied aviation medicine, specialising in the vestibular apparatus of the inner ear, which was later to prove so sensitive to weightlessness. Yegorov was one of the doctors who examined Yuri Gagarin after his flight.

Voskhod was about 1,000lb heavier than Vostok because of the two extra crewmen and their life-support requirements, and was launched by the workhorse SS-6 with an uprated second stage of a type that had been used for early planetary probes. Launched on October 12, 1964, Voskhod 1 was the only manned spaceflight of that year. Code-named *Ruby*, the mission was described as an "extended" flight and there was some surprise when it ended after just 24hr and 16 orbits. This may ·have had something to do with the downfall of Nikita Khruschev – who talked to the crew by phone during the flight but was apparently ousted in mid-sentence – though it was subsequently revealed that Feoktistov and Yegorov had been far from happy, suffering bouts of sickness. Komarov, who had been in training for four years, was not affected.

The cosmonauts carried out minor experiments and manoeuvres. They spotted a thunderstorm over Africa and reported "Glenn fireflies". They requested an extension to their flight but were ordered home by the father of Soviet manned spaceflight, Sergei Korolev himself, quoting Shakespeare: "There are more things in heaven and Earth, Horatio."

The crew's description of their re-entry was interesting. The light visible through the portholes turned pink, the black sky turned to orange-red and the space capsule was heated to 1,000°C. As Voskhod passed through the sound barrier it shuddered "like a truck on a cobbled road". Rockets fired to slow their descent just before the capsule hit the ground beneath its single parachute at a point 194 miles north-east of Kustanai after a flight of 415,936 miles.

Pavel Belyayev, commander of Voskhod 2 and the 14th man in orbit (Novosti)

Alexei Leonov practising his spacewalk inside an aeroplane (Novosti)

Name: Voskhod 2
Sequence: 19th astro-flight, 14th spaceflight, 12th Earth orbit
Launch date: March 18, 1965
Launch site: Tyuratam, USSR
Launch vehicle: A2 (SL-4)
Flight type: Earth orbit and spacewalk
Flight time: 1 day 2hr 2min 17sec
Spacecraft weight: 12,529lb
Crew: Col Pavel Ivanovich Belyayev, 39, Soviet Air Force, commander
Lt-Col Alexei Archipovich Leonov, 30, Soviet Air Force, pilot

Five days before the scheduled launch of the first manned American Gemini two-man spacecraft, the Russians stole the show yet again. They too sent up two men, but replacing the third man in the Voskhod spacecraft was a flexible telescopic airlock through which one of the spacesuited cosmonauts was to crawl out into space. So it was that the Soviet Union added the first spacewalk to its growing list of space successes.

The commander of the flight was Col Pavel Belyayev, who was born June 26, 1925, in Chelishchevo in the Vologda region, a thickly forested area north-east of Moscow. A former mill worker, he learned to fly and joined the Soviet Air Force during the war, seeing action against Japan. Belyayev was among the first group of cosmonauts, though a leg injury in 1961 delayed his space training for a year.

Voskhod 2's pilot was Alexei Leonov, born in the Altai region on May 20, 1934. He entered flying school at the age of 19 and later studied aeronautical engineering at a Soviet Air Force academy. Like Belyayev, he was a member of the original cosmonaut group.

The crew, code-named *Diamond*, blasted into the cold skies over Tyuratam at 10.00 am Moscow time on March 18, 1965, and reached a record apogee of 308 miles. On the second orbit Leonov prepared to leave the craft. He entered the extended airlock, which protruded from the Voskhod capsule by about five feet. Squeezing himself through the three-foot-wide tunnel, Leonov opened a hatch at the end and, watched by automatic cameras, eased his way out of the airlock to swim in space at the end of a 15ft-long tether.

Leonov did more than just cavort in space, turning somersaults at a rate of ten revolutions a minute. He proved that a man could work outside his craft in space and survive, given the protection of a spacesuit and a life-support backpack. The intense cold of shadow, the heat of the sun, radiation and micrometeorites were all expected to present severe problems. Leonov's exploit gave the green light for useful work in space and, ultimately, walking on the Moon.

Though his spacewalk had been a triumph, Leonov ran into severe trouble when he tried to get back into the Voskhod. His spacesuit had ballooned and it was only with great difficulty that he managed to squeeze back inside the snake-like airlock, struggling for eight minutes after his ten-minute excursion. Total time outside the pressurised cabin was 23min 41sec.

The airlock was jettisoned and Voskhod 2 continued its flight for 16 orbits. Then, as Belyayev prepared for automatic firing of the retro-rockets, a sensor in the attitude-control system failed and the automatic procedure had to be halted.

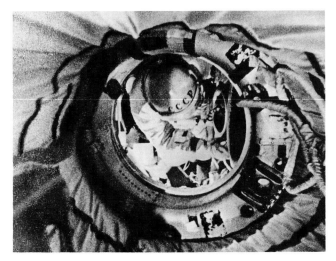

Alexei Leonov squeezes through the airlock tunnel towards open space (Novosti)

Leonov in space (Novosti)

This meant that the spacecraft had to fly on for one more orbit while Belyayev performed the orientation manoeuvres manually before firing the retros. As a result of the extra orbit Voskhod 2 landed nearly 600 miles off target in the deep snow of the forests of Perm. Flight time was 26hr 3min and distance travelled 445,420 miles.

A damaged telemetry antenna made it impossible for rescue teams to locate the craft, and it took a helicopter 2½hr to find the crew. Only then was an announcement about the end of the mission made, quashing rumours in the West about a disaster. The terrain ruled out a helicopter pick-up, however, and Belyayev and Leonov spent the night in their spacecraft hiding from wolves before being reached by ground vehicles.

The flight of Voskhod 2 proved to be Russia's final space spectacular, and America's Gemini dominated the space scene for the next two years. Belyayev succumbed to a long and serious stomach illness on January 10, 1970, becoming the first spaceman to die of natural causes.

Gemini 3   March 23, 1965   Flight 20

Name: Gemini 3 (GT-3)
Sequence: 20th astro-flight, 15th spaceflight, 13th Earth orbit
Launch date: March 23, 1965
Launch site: Pad 19, Cape Kennedy, USA
Launch vehicle: Titan II
Flight type: Earth orbit
Flight time: 4hr 52min 52sec
Spacecraft weight: 7,111lb
Crew: Maj Virgil Ivan "Gus" Grissom, 39, USAF, command pilot
Lt-Cdr John Watts Young, 35, USN pilot

Gus Grissom (left) and John Young discuss procedures before entering Gemini 3 for a communications test at Cape Kennedy on January 6, 1965 (Nasa via Astro Information Service)

The second American manned space project got under way in April 1964 with the launching of the unmanned Gemini 1 capsule on a test flight and the selection of the two crewmen for the first manned mission, Gemini 3. Project Gemini was conceived as a programme to bridge the gap between Mercury and Apollo, testing equipment, training astronauts and trying out in Earth orbit the manoeuvres needed for rendezvous and docking, the two keys to a successful lunar landing mission. The most important Apollo manoeuvre was to be the rendezvous and docking in lunar orbit of an ascent stage and an Apollo command and service module after the landing and surface exploration. Gemini 3 was to be the first spacecraft to simulate rendezvous by changing its orbital height.

The distinctive black-and-white Gemini consisted of two modules: the re-entry module, of similar configuration to but

larger than Mercury, with pressurised cabin, re-entry control, and rendezvous and recovery sections; and the adapter module, with retro-rockets and equipment. The complete spacecraft weighed about 8,000lb at launch and measured 18ft 4in long and 10ft in diameter at the base. The re-entry module was 11ft long and 7ft 6in in diameter at the base. Gemini had a 100 per cent oxygen atmosphere, an on-board guidance computer – the first in a spacecraft – 16 attitude thrusters, and a re-entry control system comprising thrusters and four solid-propellant 2,500lb-thrust retro-rockets. Later Geminis were to carry rendezvous radar and a unique fuel-cell power system.

A modified Titan ICBM was used to launch Gemini 3. The two engines of the first stage developed a total thrust of 430,000lb, and the second stage was equipped with a single engine of similar type and rated at 100,000lb. The combination was 109ft tall and 10ft in diameter. Launch site was Cape Kennedy, formerly Cape Canaveral and renamed after the assassinated president after his death in November 1963. The Gemini Titans were launched from Pad 19, a few gantries north of Mercury-Atlas Pad 14 in a row of complexes north of the tip of the Cape and in the busiest part of the spaceport.

The Gemini 3 crew were chosen on April 13, 1964. Command pilot Virgil Grissom, carrying out his first orbital mission, would be the first man to make two spaceflights. Pilot John Young was born on September 24, 1930, in San Francisco and gained an aeronautical engineering degree. He joined the US Navy in 1952, becoming a test pilot in 1959. In 1962, the year of his selection among the second group of Nasa astronauts, Young set world time-to-height records at 3,000m and 25,000m in the F-4 Phantom fighter.

Grissom and Young entered their capsule early in the morning on March 23, 1965, and were launched into space at 9.24 am, 24min later than scheduled. The humour of Grissom's choice of name for his Gemini – *Molly Brown*, after the unsinkable lady of that name in a broadway musical – was lost on Nasa and the spacecraft was not officially named. Even so, the capcom at the Kennedy control centre, Gordon Cooper, shouted as the Titan left the pad: "You're on your way, *Molly Brown!*"

With the aid of their computer the astronauts were soon performing orbit-change manoeuvres, the last of which took Gemini 3 to a low point of 52 miles. The short, three-orbit, shakedown flight ended with a splashdown in the Atlantic 59 miles off target, some nine miles from the coastguard cutter *Diligence* near Grand Turk Island. During the wait for the prime recovery ship, USS *Intrepid*, Grissom was seasick. "Gemini may be a good spacecraft but she's a lousy ship," he reported. There were two other episodes at the end of an otherwise very successful flight. Adoption of a new descent mode meant that the spacecraft was falling in a horizontal attitude when the parachute opened. The resulting whiplash smashed Grissom's faceplate into the control panel. Then, after splashdown, a heat build-up within the capsule caused the astronauts to abandon Gemini 3 and their spacesuits. The following day's front pages pictured them in their underwear as they made a rather inglorious walk along the traditional red carpet aboard the recovery ship. According to John Young, however, the only thing really wrong with the three-orbit mission was the fact that it didn't last long enough.

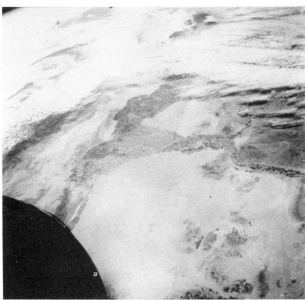

Top *A Titan 2 lifts off from Pad 19 to inaugurate the manned Gemini programme.* **Left** *A view of the California-Mexico border from Young's window. The nose of Gemini 3 can be seen at bottom left* (Nasa)

Name: Gemini 4 (GT-4)
Sequence: 21st astro-flight, 16th spaceflight, 14th Earth orbit
Launch date: June 3, 1965
Launch site: Pad 19, Cape Kennedy, USA
Launch vehicle: Titan II
Flight type: Earth orbit and spacewalk
Flight time: 4 days 1hr 56min 12sec
Spacecraft weight: 7,879lb
Crew: Maj James Alton McDivitt, 35, USAF, command pilot
Maj Edward Higgins White II, 34, USAF, pilot

There is little doubt that as a result of the Voskhod 2 spacewalk the Americans speeded up their own plans for "extravehicular activity" (EVA). Gemini 4 was originally scheduled to be a four-day flight involving in-orbit manoeuvring, with just a possibility that the pilot would open the hatch on his seat to perform a "stand-up EVA". But by May 1965 Nasa had announced that not only would Gemini 4 perform stationkeeping manoeuvres with the Titan II second stage but that a full EVA would take place. This ambitious schedule was a reflection of the fact that at last the Americans believed themselves ready to catch up with the Russians. The two pilots of Gemini 4 – which was going to be called *Little Eva*, *American Eagle* or *Phoenix* before Nasa's clampdown on spacecraft names – were both USAF majors and both from the second group of Nasa astronauts, chosen in September 1962.

Command pilot James McDivitt was born in Chicago on June 10, 1929, and first worked as a plumber. He joined the USAF in 1951, flying 145 combat sorties in Korea and winning the DFC and five Air Medals. He subsequently studied aeronautical engineering, graduating first in his class, attended the test pilot and aerospace research pilot schools at Edwards Air Force Base, and had been chosen to fly the X-15 rocket plane when his selection by Nasa was announced.

Gemini 4 pilot Edward White came from San Antonio, Texas, where he was born on November 14, 1930. He received a science degree from West Point in 1952 and joined the USAF, flying in Germany and as a test pilot in the US. He narrowly missed selection as one of the Mercury team, and at that time one of his tasks was to fly large transport aircraft on parabolic paths, simulating zero g for astronaut trainees. He graduated with a degree in aeronautical engineering in 1959.

The launch of Gemini 4 at 10.15 am on June 3, 1965, was delayed 76min and was the first to be shown live in Western Europe, via the Early Bird comsat. Gemini 4 was also the first mission to be controlled from the new Manned Spacecraft Centre in Houston, Texas.

Once in orbit, McDivitt's first job was to fly close to and in formation with the second stage of the Titan II. This proved altogether more difficult than in simulations, and after using up 42 per cent of the manoeuvring fuel in the first few minutes of the scheduled four-day flight McDivitt was ordered to curtail the experiment. "Knock it off. No more rendezvousing with the booster," said capcom Virgil Grissom.

The EVA was postponed for one orbit. Then, on the third pass around the Earth, Edward White opened his hatch and floated into space. With the aid of a hand-held manoeuvring unit White gambolled weightlessly and formed the subject of some of the finest space photographs ever taken. Conversation during the EVA was more chatty than scientific and, while the whole world listened in, ground control could hardly get a word in edgeways. Grissom had to call Gemini 4 many times to order White back into the craft before it

Top *Ed White, left, and Jim McDivitt inside Gemini 4 before launch.*
Left *Ed White during his EVA* (Nasa)

entered the Earth's shadow. The conversation went like this:
White: "The manoeuvring unit is good. The only problem I have is I haven't enough fuel ... There is absolutely no disorientation."
McDivitt: "Get out in front where I can see you ... Where are you? Move slowly and I'll take your picture ... Ed, will you please roll around ... Hey Ed, smile ... You smeared up my windshield, you dirty dog! ... Gus, this is Jim, got any messages for us?"
[After several attempts by Grissom to contact the spacecraft:]
Grissom: "Get back in!"
McDivitt: "OK, they want you to get back in now."
White: "This is fun!"
McDivitt "Well, back in, come on."
White: "I'm coming. Aren't you going to hold my hand?"
McDivitt: "No ... come on. Let's get back in here before it gets dark".
White: "This is the saddest moment of my life."
White's EVA lasted 21min and the hatch open-to-close time

was 36min. His heart rate started at 150 and peaked at 178. The astronauts had great difficulty in closing the hatch, but after this excitement the rest of the flight was something of an anticlimax, although it was given mammoth press coverage because it clocked up a US record of 62 orbits. The astronauts also conducted 11 scientific experiments. Then, approaching the end of the mission, the computer failed and McDivitt was forced to perform a manual re-entry. He reduced the orbit from approximately 100-175 to 48-99 miles and fired the retros, initiating an 8g re-entry and jettisoning the adapter module. Gemini 4 came down 390 miles east of the Cape, 48 miles off target, and the two astronauts had to wait a while before the helicopters from the USS *Wasp* came to pick them up.

As he was being examined in the sick bay McDivitt quipped: "I always knew I would end up in hospital after four days in space". White contented himself with a simple "Yahoo!" Both astronauts knew that their 1,609,700-mile journey in 97hr 56min had put America on an equal footing with the Russians. It was left to Gemini 5 to take the lead.

X-15 No 3, Flight No 44
X-15 No 3, Flight No 46

June 29, 1965
August 10, 1965

Flight 22
Flight 23

Name: X-15 No 3, Flights No 44 and 46
Sequence: 22nd and 23rd astro-flights
Flight dates: June 29, 1965; August 10, 1965
Take-off site: Edwards Air Force Base, California, USA
Launch vehicle: B-52 carrier aircraft
Flight type: Sub-orbital
Flight times: 10min 32sec, 9min 51sec
Spacecraft weight: About 34,000lb
Crew: Capt Joseph Henry Engle, 32, USAF

Joe Engle was born in Abilene, Kansas, on August 26, 1932. He graduated from the University of Kansas with a degree in aeronautical engineering before entering the Air Force in 1955. He was top student in his flying classes, took up a gunnery course and was assigned to squadrons in the US and Europe, including Italy, Denmark and Spain. In 1961 he graduated from the USAF Experimental Test Pilot School and then studied at the Aerospace Research Pilot School.

Engle was selected to fly the X-15 rocket aircraft in 1963 and made his first flight on October 7 of that year. In 1964, when he flew seven X-15 missions, Engle was chosen as one of the "Nation's Ten Outstanding Young Men" by the Junior Chamber of Commerce and as the "Outstanding Young Air Force Officer of 1964" by the USAF Association.

On June 29, 1965, Engle climbed aboard X-15 No 3 for its 44th flight, his 14th and the programme's 138th. During the high-altitude mission he was to conduct a horizon-scanning experiment, among other tasks. Released over Delamar Lake, he reached a speed of 3,432mph (Mach 4.94) and a height of 280,600ft (53.14 miles) after an 81sec burn. The flight lasted 10½min.

On August 10, during his next mission, the programme's 143rd and X-15 No 3's 46th, Engle reached a height of 271,000ft (51.7 miles) following an 82sec burn, making him the third man to fly above 50 miles twice. Maximum speed for the 9.86min flight was 3,550mph (Mach 5.20).

Joe Engle pictured on joining the X-15 programme in 1963, when he was 30 years old and a test pilot at Edwards AFB (Nasa via Astro Information Service)

Name: Gemini 5 (GT-5)
Sequence: 24th astro-flight, 17th spaceflight, 15th Earth orbit
Launch date: August 21, 1965
Launch site: Pad 19, Cape Kennedy, USA
Launch vehicle: Titan II
Flight type: Earth orbit
Flight time: 7 days 22hr 55min 14sec
Spacecraft weight: 7,949lb
Crew: Lt-Col Leroy Gordon Cooper, 38, USAF, command pilot
Lt-Cdr Charles "Pete" Conrad Jr, 35, USN, pilot

Pete Conrad (left) and Gordon Cooper in good spirits before their marathon Gemini 5 flight (Nasa)

A sleepless Conrad pictured inside Gemini 5 during what he described as a very boring spaceflight. An attitude-control system failure reduced the mission to long periods of drifting flight (Nasa via Astro Information Service)

The motto for the flight of Gemini 5 was "Eight days or bust". The badge designed for the two spacemen's suits depicted a wagon of the Western pioneering days, when the slogan was "California or bust". The spacemen ran into a few ambushes during their trip, in the form of a power failure, malfunctioning thrusters and a hurricane which forced them home one orbit short of the full eight days. They were Mercury veteran Gordon Cooper and Charles Conrad, an irrepressible little test pilot from the second group of astronauts.

Charles Conrad was born in Philadelphia on June 2, 1930, and earned a degree in aeronautical engineering at Princetown University in 1953 before entering the Navy. He attended test pilot school at Patuxent River, Maryland, and served as a flight instructor, project test pilot, performance engineer and safety officer in various US Navy units before his selection as an astronaut. Like Edward White of Gemini 4, he narrowly missed selection for Mercury.

Launch of Gemini 5 was postponed 48hr to 9.00 am on August 21, when the spacecraft achieved a US manned spaceflight first by departing right on schedule. The 7,949lb Gemini 5 continued on its winning way, entering an orbit which took the astronauts to a US record height of 217 miles. Another first was the recovery of the Titan's first stage in the Atlantic.

Two hours into the mission Cooper, the first man to enter orbit twice, was to eject a 76lb radar evaluation pod (REP) from the Gemini's nose, back away 52 miles and then rendezvous with the REP. This manoeuvre was designed to test the guidance and navigation systems being carried aboard for the first time. Also being carried for the first time was the fuel-cell power system, which converted oxygen and hydrogen into electrical power, creating water as a by-product.

The REP trial had to be cancelled when the oxygen pressure in the fuel cell system began to decrease towards a critical level. The pressure dropped from 800lb/in² to 150, then 140 and 120, where it seemed to hold steady for a while. The crew was ordered to switch off everything possible to conserve power, aircraft were dispatched to an emergency splashdown zone in the Pacific, and it looked as though it would be a case of "three orbits and bust" for the astronauts. Unltimately the pressure dropped to 60lb/in², but by then the astronauts had decided to try for a day in space at low power. The failure was the result of a fault in a heater, which, inexplicably, returned to normal the following day. The pressure rose and, after a day of drifting, tumbling, chatting, eating, singing ("Over the ocean, over the blue, here's Gemini 5 singing to you") and looking out of the window, the

astronauts were given the go-ahead to tackle a limited trials programme.

With a certain amount of manoeuvring now allowed, Gemini 5 made a rendezvous with a "phantom" target, making five orbit changes during a two-hour, 40,000-mile chase. The crew carried out 17 science experiments, one of which examined man's ability to spy from space. The astronauts failed to see a number of specially constructed targets on the ground but did spot a rocket lifting off from Vandenberg AFB, California, and the wake of their prime recovery ship, the USS *Lake Champlain*, waiting for them in the Atlantic Ocean. Later six attitude-control thrusters failed and the astronauts were reduced to tumbling and sightseeing again. When they finally exceeded the Vostok 5 record of 119hr 6min, all Cooper could say was: "At last, huh?" In the latter stages of the mission the astronauts spoke to Scott Carpenter, who was working 62m beneath the Pacific in Sealab 2.

During their final orbits, while travelling over South America, the astronauts spied Hurricane Betsy in the Atlantic, heading straight for their splashdown point. Mission Control decided to change the landing site by ending the mission one orbit early, but incorrect co-ordinates were fed into the craft's computer and the astronauts landed 90 miles short of the planned point 335 miles south-west of Bermuda. They had covered over three million miles and 120 revolutions in seven days 22hr 55min. One newspaper summed up the flight: "Man has got what it takes to fly to the Moon and back. Now for the Moon shot."

X-15 No 3, Flight No 49   September 28, 1965   Flight 25

Name: X-15 No 3, Flight No 49
Sequence: 25th astro-flight
Flight date: September 28, 1965
Take-off site: Edwards Air Force Base, California, USA
Launch vehicle: B-52 carrier aircraft
Flight type: Sub-orbital
Flight time: 11min 56sec
Spacecraft weight: About 34,000lb
Crew: John B. "Jack" McKay, 42

The objectives of the 150th X-15 mission were to investigate boundary-layer noise and structural loads on the horizontal tail, and to carry out horizon-scanning experiments. The flight took place on September 28, 1965, and was the 49th by the third X-15. Maximum speed was 3,732mph (Mach 5.33) and maximum altitude 295,600ft (56 miles), making pilot Jack

Jack McKay, X-15 pilot and 26th man above 50 miles (Nasa)

McKay an unofficial astronaut. Burn time was 80.8sec and the flight lasted 11min 57sec.

McKay was born in Portsmouth, Virginia, on December 8, 1922, and flew for the US Navy in the Second World War, earning the Air Medal with two clusters and a Presidential Unit Citation. He gained a science degree in 1950 and joined NACA in 1951 to fly the Skyrocket and X-1 experimental rocket aircraft, and to serve as project pilot on the F-100, 102, 104 and 107 programmes. Becoming the third civilian X-15 pilot on October 28, 1960, he carried out one mission in 1961. The following year he flew five, the last of which, in X-15 No 2, almost killed him when the vehicle broke up during an emergency landing. He was gravely injured and lay trapped in the wreckage for many hours.

Remarkably, though he never fully recovered from his back injuries, McKay survived to make 13 more flights in the X-15 before his astro-flight. McKay died of cancer in April 1975, aged 52. He left eight children, the most fathered by any spaceman.

Name: X-15 No 1, Flight No 61
Sequence: 26th astro-flight
Flight date: October 14, 1965
Take-off site: Edwards Air Force Base, California, USA
Launch vehicle: B-52 carrier aircraft
Flight type: Sub-orbital
Flight time: 9min 18sec
Spacecraft weight: About 34,000lb
Crew: Capt Joseph Henry Engle, 33, USAF

Neil Armstrong flew the X-15 rocket aircraft and became the first man to walk on the Moon. Joe Engle emulated Armstrong to the point of becoming a Nasa astronaut and being tentatively scheduled for an Apollo lunar landing, but he never actually got to the Moon.

On his 14th and 15th X-15 missions earlier in 1965 Engle had exceeded 50 miles altitude, becoming an unofficial astronaut. He made it three in a row on his 16th flight. This one, the 153rd of the programme and the 61st by X-15 No 1, was made to carry out measurements of atmospheric pressure and to perform horizon-scanning experiments. Engle attained 3,554mph (Mach 5.08) and 266,500ft (50.17 miles). Burn time was 84.8sec and the flight lasted 9min 18sec. The mission was Engle's final X-15 flight before moving to Houston and an astronaut career which, after a long, patience-testing wait, culminated in command of one of the early Shuttle test flights.

Joe Engle pictured in 1969, when he was a member of the Apollo 10 support crew (Nasa via Astro Information Service)

James Lovell, left, and Frank Borman after their 13-day flight aboard Gemini 7 (Nasa)

Name: Gemini 7 (GT-7)
Sequence: 27th astro-flight, 18th spaceflight, 16th Earth orbit
Launch date: December 4, 1965
Launch site: Pad 19, Cape Kennedy, USA
Launch vehicle: Titan II
Flight type: Earth orbit and rendezvous
Flight time: 13 days 18hr 35min 1sec
Spacecraft weight: 8,076lb
Crew: Lt-Col Frank Borman, 37, USAF, command pilot
Lt-Cdr James Arthur Lovell Jr, 37, USN, pilot

Name: Gemini 6 (GT-6A)
Sequence: 28th astro-flight, 19th spaceflight, 17th Earth orbit
Launch date: December 15, 1965
Launch site: Pad 19, Cape Kennedy, USA
Launch vehicle: Titan II
Flight type: Earth orbit and rendezvous
Flight time: 1 day 1hr 51min 54sec
Spacecraft weight: 7,817lb
Crew: Capt Walter Marty Schirra Jr, 42, USN, command pilot
Maj Thomas Patten Stafford Jr, 35, USAF, pilot

The greatest hurdle on the American road to the Moon was rendezvous and docking. Proving the necessary techniques was to be the task of Gemini 6. On October 25, 1965, an Agena target vehicle lifted off from Pad 14 at Cape Kennedy atop an Atlas launcher. Gemini 6 was to be launched later to rendezvous and dock with it. It was not to be: the Agena exploded, leaving Gemini 6 crew Wally Schirra and Tom Stafford with nowhere to go except back to their quarters.

Stafford was born in Weatherford, Oklahoma, on September 17, 1930, and graduated from the US Naval Academy in 1952 before entering the USAF. He flew fighters and attended the USAF Experimental Test Pilot School. Following graduation he served as chief of the performance branch of the Aerospace Research Pilot School. He joined the second class of Nasa astronauts in 1962 and first teamed up with Schirra on the Gemini 3 back-up crew.

The flight of Gemini 7, on a 14-day extended mission packed with 20 science experiments, was scheduled for December 1965. Seeking to recover lost ground, Nasa took the bold step of planning a dual rendezvous mission. On October 28 President Johnson announced that Gemini 7 would be launched first and would act as the rendezvous target for Gemini 6, setting the stage for the most exciting spaceflight so far. The only question was whether Pad 19 could be repaired and prepared for a launch so soon after Gemini 7's ascent.

The Gemini 7 crewmen, Gemini 4 back-ups Frank Borman and James Lovell, were to endure 14 days of weightlessness in a space no larger than the front seat of a car. They were the nearest one could get to space twins: both were 37, both were born in March 1928, both had blonde hair and blue eyes, and both were chosen in the second group of Nasa astronauts.

Born on March 14, 1928, in Gary, Indiana, Frank Borman was raised in Tucson, Arizona. In 1950 he graduated from the US Military Academy with a science degree. After flight training in the USAF he returned to the Academy to teach thermodynamics and fluid mechanics and later studied aeronautical engineering at the University of California. He was one of the first graduates of the USAF Aerospace Research Pilot School at Edwards Air Force Base, and before his selection as an astronaut he served as an instructor there.

Jim Lovell, a lieutenant-commander in the US Navy, was one of the few astronauts to have a real nickname, his nervous energy and bouncy character earning him the sobriquet "Shaky". He was born in Cleveland, Ohio, on March 25, 1928. From 1946 to 1952 he attended the University of Wisconsin and the US Naval Academy. Following flight training he served in various squadrons until 1958, when he became a test pilot at Patuxent River. In 1961 he graduated in flight safety from the University of Southern California.

Apart from its other achievements, Gemini 7 might have claimed the honour of being the first spacecraft to be launched from a football pitch. Because manned flights no longer seemed to warrant full live coverage, a US television station compromised by showing the launch, at 2.30 pm, on the corner of the screen while a football match was being covered. Borman and Lovell overcame early fuel-cell problems and quietly clocked up seven days in space. "I feel as though I was born here," said Borman. In a first of sorts, Lovell had removed his spacesuit and was flying in his underwear. Living in the confines of the spacecraft was not without its difficulties, however. When a urine bag came apart in Borman's hands he was asked: "Before or after?" "After," he replied ruefully.

Gemini 6 crew Tom Stafford, left, and Wally Schirra (Nasa via Astro Information Service)

December 12, 1965, was the big day. The Gemini 6 countdown reached zero and, with a familiar high-pitched whine, the Titan II's engines started up. Aboard Gemini 7, orbiting above Cape Kennedy, Borman shouted, "I see ignition .. Oh-oh, I see shutdown." A plug had fallen out of a malfunction detection system on the Titan and the burn had been halted after 1.2sec. With hypergolic fuel still in the lines, there was a danger that the Titan could explode. Inside Gemini 6 the countdown clock had started, but Schirra knew that there had been no lift-off. Afterwards the astronauts were praised for not taking to the launch escape system, as mission rules demanded. Their coolness saved the launch and they were ready to go again at 8.37 am on December 15.

This time things went perfectly. Three orbits later, Gemini 6 rendezvoused with Gemini 7. This apparently simple feat had taken seven carefully calculated orbital manoeuvres: at

Gemini 7 seen from the approaching Gemini 6 (Nasa)

perigee a forward burn to raise apogee nine miles; at second apogee a forward burn to raise perigee 40 miles; burn to match Gemini 7 orbital plane; tweak burn to raise apogee half a mile; at third apogee a forward circularisation burn; rendezvous begins with forward burn to transfer to Gemini 7 orbit; forward burn to match Gemini 7 velocity.

"We did it!" shouted Stafford, and as the two craft came within six inches of each other the crews exchanged lighthearted banter. Theirs was the greatest space achievement since Gagarin. Five hours 18min later the Geminis parted. The Gemini 6 crew, playing "Jingle Bells" and pretending to be a UFO called Santa Claus, made a perfect manual re-entry, splashing down seven miles from USS

Wasp, 630 miles south-west of Bermuda. Their flight had lasted 25hr 51min and 16 revolutions.

Borman and Lovell continued their endurance flight, proving beyond a doubt that man could live and work in space for two weeks. They were in good but stooping condition as they walked across the deck of the *Wasp* after a pinpoint splashdown 700 miles south-west of Bermuda and just 6.5 miles away from the ship. Gemini 7 had set remarkable new records for duration and distance travelled: 330hr 35min, 206 orbits and 5,716,900 miles.

"You have moved us one step higher on the stairway to the Moon," President Johnson told the elated astronaut quartet after the mission that must surely rank as one of the most important in the history of space exploration.

Gemini 8   March 16, 1966   Flight 29

Name: Gemini 8 (GTA-8)
Sequence: 29th astro-flight, 20th spaceflight, 18th Earth orbit
Launch date: March 16, 1966
Launch site: Pad 19, Cape Kennedy, USA
Launch vehicle: Titan II
Flight type: Earth orbit, rendezvous and docking
Flight time: 10hr 41min 26sec
Spacecraft weight: 8,351lb
Crew: Neil Alden Armstrong, 35, command pilot
Major David Randolph Scott, 34, USAF, pilot

The first serious space emergency came at a time when America at least had come to expect effortless success, and it demonstrated the dangers of spaceflight to a nation that had become complacent. Gemini 8 was to perform the first space docking, with a 7,000lb Agena target vehicle. Also on the schedule was an EVA lasting two hours, during which the pilot would walk around the world.

Commander of Gemini 8 was civilian test pilot Neil Armstrong and pilot was David Scott, a USAF major. Armstrong was born in Wapakoneta, Ohio, on August 5, 1930, and learned to fly before he could drive. He became a US Navy pilot in 1949, flew 78 combat missions over Korea, and was shot down and rescued by helicopter. At the end of the Korean War Armstrong joined NACA (later to become Nasa) and participated in flight-test work on the F-100, F-102, F-104 and B-47. Later he was chosen for the X-15 rocket aircraft, which he first flew on November 30, 1960. By the time he had joined the second group of Nasa astronauts in 1962 he had flown to a height of 40 miles and at a speed of Mach 5.75. His first astronaut assignment was as Gemini 5 back-up command pilot.

David Scott was the first of the Group 3 Nasa astronauts, chosen in October 1963, to make a spaceflight. He was born in San Antonio, Texas, on June 6, 1932, and graduated in science from the US Military Academy and in aeronautics and astronautics from Massachusetts Institute of Technology (MIT). Scott was also a graduate of the Experimental Test Pilot School and Aerospace Research Pilot School at Edwards Air Force Base and in 1963 survived a serious jet crash (with X-15 pilot Michael Adams).

The Gemini 8 launch was postponed for a day when trouble developed in the Atlas booster of the Agena 8 target

rocket. On March 16, 1966, first the Agena and then, at 11.41 am, Gemini 8 were launched into Earth orbit. After manoeuvring for about 4hr 40min, simulating the ascent of a lunar module from the Moon's surface, the astronauts sighted the Agena. Scott could see the target shining like a star 76 miles away. "We have some object in sight. It looks like the Agena. We've got a real winner here," he reported.

About an hour and a half later the astronauts had moved to within 50ft of the target. Using the Gemini thrusters,

Neil Armstrong (left) and David Scott pose at Cape Kennedy before their Gemini 8 mission (Nasa via Astro Information Service)

Armstrong moved ahead of the Agena and turned around so that the spacecraft's nose was facing the docking collar on the target. He then moved in slowly, gingerly jockeying Gemini 8 into position for docking. When controllers on the South Pacific tracking ship USS *Rose Knot Victor* told the spacemen to go ahead and dock they were surprised by Armstrong's elated replay: "We are docked ... and she's a real smoothie!"

The first docking in space had been achieved some six and a half hours after launch. Preparing to manoeuvre the combined spacecraft, the astronauts passed commands into the Agena's control system, but then trouble began. As the combination coasted around the Earth 27min after docking, it went into a spin. Armstrong and Scott carefully weighed the situation and decided that the fault must be with the Agena, which was probably firing one of its thrusters uncommanded.

"We're backing off," said Armstrong. The craft separated but the trouble got worse. Though the Agena steadied, the manned craft continued to rotate at up to 360° a second, making a catastrophic collision a possibility. Armstrong exclaimed: "We've got serious problems here ... we're tumbling end over end ... and we can't turn anything off!" More diagnosis by the astronauts revealed that the only way to control the spin was to use the re-entry orientation system. Armstrong deactivated the other controls and used up 75 per cent of the RCS fuel before regaining control of Gemini 8 mid-way through the seventh orbit.

Under mission rules the low fuel state demanded an emergency return to Earth. While a tense world listened in, Armstrong fired the retros over Southern China and the crew splashed down 1.1 miles from their target, 690 miles southeast of Okinawa in the Pacific. Flight time was 10hr 41min and Gemini 8 had travelled 181,500 miles. A recovery pilot from *Leonard F. Mason* sighted the craft shortly afterwards and mistakenly reported that the "doors were open and they were just sitting there enjoying the sun and eating lunch." In fact the space duo had been badly seasick during their three-hour wait, the longest in the Gemini programme.

Post-flight investigation revealed that the emergency had resulted from a short circuit in the Gemini control system which caused one thruster to fire continuously.

Gemini 8 approaches the Agena 8 target rocket (Nasa via Astro Information Service)

Below *Launch of Gemini 8 by a Titan II.* **Bottom** *Gemini 8's nose enters the Agena docking collar* (Nasa)

Name: Gemini 9 (GTA-9)
Sequence: 30th astro-flight, 21st spaceflight, 19th Earth orbit
Launch date: June 3, 1966
Launch site: Pad 19, Cape Kennedy, USA
Launch vehicle: Titan II
Flight type: Earth orbit, rendezvous and spacewalk
Flight time: 3 days 0hr 20min 50sec
Spacecraft weight: 8,087lb
Crew: Lt-Col Thomas Patten Stafford Jr, 35, USAF, command pilot
Lt-Cdr Eugene Andrew Cernan, 32, USN, pilot

Gemini 9 crew Tom Stafford (left) and Gene Cernan (Nasa via Astro Information Service)

On February 28, 1966, two T-38 jet trainers flew over the city of St Louis. On board were the prime and back-up crewmen of the Gemini 9 mission, scheduled to take place later in the year. The astronauts were on their way to see the Gemini 9 spacecraft being built at the McDonnell Douglas factory in St Louis. Visibility was very poor, and the first aircraft made an unsuccessful attempt to land at the airport adjoining the factory. As it came in a second time the T-38 hit the roof of the building housing Gemini 9, bounced off, smashed into a car park and exploded. The two men on board, Gemini 9 prime crew Elliot See and Charles Bassett, were killed. See was a Group 2 astronaut and Bassett was from Group 3.

Aboard the second aircraft, which made a safe landing, were the back-up crew, Lt-Col Tom Stafford and Lt-Cdr Eugene Cernan, who thus became the first reserves to take over an American space mission since Scott Carpenter.

Cernan was born in Chicago, Illinois, on March 14, 1934, and studied electrical engineering at Purdue University before becoming a US Navy pilot in 1956. He began a degree course in aeronautical engineering in 1961 and had graduated by the time of his selection with the third group of Nasa astronauts in October 1963. Aged 32 at the time of the flight, Cernan was the youngest American ever to fly in space.

The flightplan for Gemini was basically the same as that of its predecessor: a rendezvous and docking with an Agena target vehicle, a firing of the Agena's main engine to take the docked combination into a higher orbit, a number of scientific experiments, and a spacewalk by Cernan using the USAF-developed Astronaut Manoeuvring Unit (AMU), which would allow him to "fly" about. The AMU, a backpack into which Cernan would strap himself, was nicknamed the "flying armchair" by the astronauts.

The ambitious flight had got off to a sad beginning with the deaths of See and Bassett and proceeded in catastrophic fashion with the explosion of the Agena target rocket during its launching from Pad 14 at Cape Kennedy on May 17. This was the second failure in three attempts to get an Agena off the ground, but this time Nasa had prepared for such an eventuality by building a secondary, less versatile, target called the Augmented Target Docking Adapter (ATDA), which was quickly prepared and duly launched by an Atlas on June 1. Two miles away, on top of their Titan II on Pad 19, Stafford and Cernan entered the final two minutes before their launch. Then, updated launch data were rejected by the Titan's control-system computer and the flight had to be cancelled. This was the fourth time that Stafford had been left

on the launch pad in five attempts to get off the ground.

While waiting 48hr for a new launch attempt, which successfully took place on June 3, mission controllers studied signals from the ATDA indicating that the two shrouds fitted to protect the docking collar had failed to jettison. No-one could be certain of the situation until the crew saw it for themselves.

After launch at 8.39 am Gemini 9 entered an initial 99/166-mile orbit and the crew first sighted the ATDA at T+3hr 20min, when it was about 60 miles away. After complicated orbit and speed-change manoeuvres lasting about three hours Stafford confirmed the controllers' suspicions with the words that were to become the Gemini 9 trademark. The ATDA, he said, "looks like an angry alligator." The shrouds had partially opened but had not been ejected, and there were a number of suggestions about how to get them off so that the docking could go ahead. Stafford himself wanted to try and nudge them off with the nose of the Gemini. Indeed, he almost came within touching distance during close manoeuvres around it. Finally the docking was called off and two more rendezvous exercises were scheduled instead. The Gemini backed away from the ATDA after a 46min rendezvous and returned from orbits higher and lower than the target for rendezvous lasting 39 and 77min respectively.

Stafford and Cernan were so tired after these experiments that the pilot's spacewalk was put back for one day. When Cernan did eventually emerge from the security of the cabin at about T+46hr the troubles that had plagued Gemini 9 returned yet again. Cernan had to work so hard during his spacewalk – trying to stay in one place and not float away at the slightest touch against the spacecraft – that his spacesuit's environmental control system could not cope with the body heat that he generated. Things got really bad when Cernan managed to get to the rear of Gemini 9 and sit in the "flying armchair". His visor had almost completely fogged over. "I can see where my nose is," he said breathlessly, "but not where my eyeballs are." He was exhausted and, to make matters worse, communications between him and his commander became impossible when he plugged into the

Left *The "angry alligator" target vehicle, showing the jammed payload shrouds that prevented docking.* **Right** *Gene Cernan pictured during his record-breaking spacewalk* (Nasa)

AMU circuit. Stafford called the spacewalker back in. The EVA had lasted a record 2hr 8min but had posed more problems than it had solved.

The rest of the flight, which included seven science experiments, went well. The splashdown was seen by millions of television viewers as Gemini 9 came down less than a mile from the prime recovery ship, USS *Wasp*, in the Atlantic

Ocean 345 miles east of Bermuda. This was one of the better moments of the flight. The crew stayed in the capsule until it had been hauled aboard and when they alighted they headed to a telephone for the now customary post-flight conversation with President Johnson. "It was a two-man job all the way," said Stafford. The two-man job had lasted three days and 21min, 44 revolutions of the Earth and 1,255,630 miles.

Gemini 10   July 18, 1966   Flight 31

John Young (left) and Mike Collins walk the red carpet after their Gemini 10 flight (Nasa via Astro Information Service)

Name: Gemini 10 (GTA-10)
Sequence: 31st astro-flight, 22nd spaceflight, 20th Earth orbit
Launch date: July 18, 1966
Launch site: Pad 19, Cape Kennedy, USA
Launch vehicle: Titan II
Flight type: Earth orbit, rendezvous, docking, re-boost and spacewalk
Flight time: 2 days 22hr 46min 39sec
Spacecraft weight: 8,295lb
Crew: Cdr John Watts Young, 36, USN, command pilot
Maj Michael Collins, 36, USAF, pilot

At the time when its two crewmen were selected earlier in the year Gemini 10 was scheduled for launching on July 18, 1966. There was not one subsequent delay and in the course of this highly successful mission the astronauts performed a space rendezvous and docking, a "re-boost" trip into high Earth orbit, a rendezvous with a second target and a spacewalk. The command pilot was Gemini 3 pilot and Gemini 6 back-up pilot Cdr John Young. The pilot was USAF major Michael

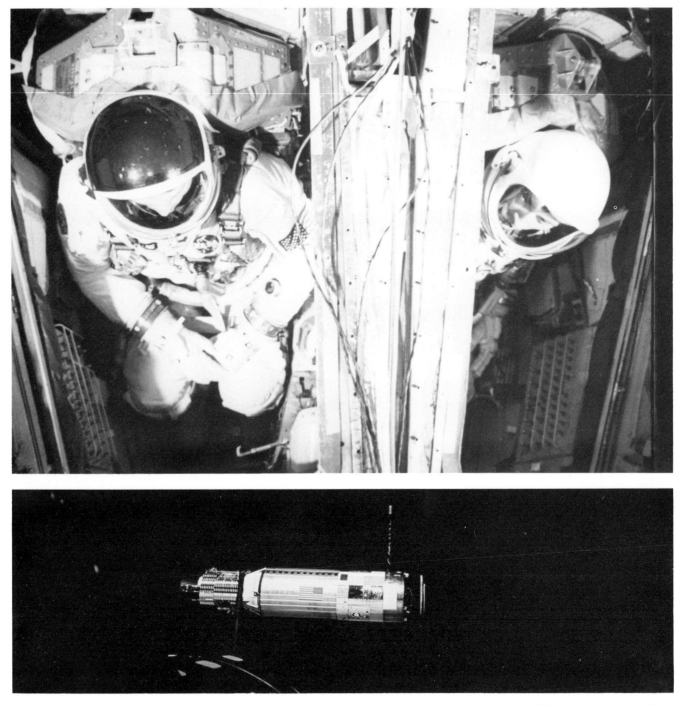

Top Inside the cramped quarters of Gemini 10 before the hatch is closed are Young (right) and Collins, with dark EVA visor on his helmet (Nasa via Astro Information Service). **Above** *Gemini 10 target rocket photographed before the docking* (Nasa)

Collins, chosen with the Group 3 astronauts in October 1963.

Born in Rome on October 31, 1930, Collins attended the US Military Academy, graduating with a science degree in 1952. He was a test pilot at Edwards Air Force Base, evaluating aircraft performance, stability and control characteristics, and served as Gemini 7 back-up pilot.

Ninety minutes before the Gemini 10 launch an Atlas Agena lifted off from Pad 14 to send the Agena 10 target into orbit. Meanwhile, Young and Collins were aboard Gemini 10,

checking systems and awaiting their own departure. The launch, at 5.20 pm, went smoothly, with few comments from the two taciturn astronauts.

The first job on the agenda was a rendezvous with the Agena 10 target. This was accomplished on the Gemini's third orbit – though not without an excessive use of fuel by Young – and docking followed shortly afterwards. The Agena was then commanded to re-start its 16,000lb-thrust engine and, for the first time ever, take the Gemini/Agena combination into a higher orbit. "It may be only 1g," said Young of the acceleration, "but it's the biggest 1g we ever saw." The resulting orbit, with its record apogee of 474 miles, was held for two revolutions before being lowered in preparation for a rendezvous with Agena 8.

During the time between docking and the high-altitude boost Collins had opened his hatch, stood on the seat and performed a stand-up EVA lasting 49min. During this time he took photographs of the Earth and collected strips of "stardust" collector that had been fixed to the outside of the craft before launch. The EVA was cut short when acrid fumes in the breathing system irritated the astronauts' throats and eyes. The fumes were later found to be lithium hydroxide, a chemical component of the environmental control system that converted the astronauts' exhaled breath into oxygen.

Leaving Agena 10 in orbit after 38hr 47min docked, Young and Collins went in search of Agena 8, which had been in orbit since March 16. Perfect manoeuvring led to a successful rendezvous, achieved entirely without recourse to radar. Then, while Young kept the spacecraft steady, Collins took a spacewalk. "Mike's going outside now," said Young as

Collins floated outside the craft, which was orbiting in formation with the Agena. Collins "swam" over to the Agena 8, using a hand-held manoeuvring unit and becoming the first person to make bodily contact with another orbiting spacecraft. Collins retrieved another stardust collection package from the side of the rocket stage, but lost his camera while trying to control his movements.

It soon however became clear that Young was using too much of the dwindling fuel supply in keeping the craft stabilised during the walk, and he reluctantly ordered Collins back inside Gemini 10 after only 39min of the planned 1½hr walk. Unfortunately, a camera fault meant that no photographs of the EVA were produced. Later the crew performed 13 science experiments before preparing to come home. "We had a lot of fun," said Collins afterwards.

On July 21 a pinpoint splashdown in the Atlantic 3.4 miles from USS *Guadalcanal* and 529 miles east of the Cape ended a successful flight that had been spoiled only by Collins' curtailed EVA and Young's overuse of fuel during the initial stages of the first rendezvous with Agena 10. The flight lasted 70hr 46min, 43 revolutions and 1,223,370 miles.

Gemini 11	September 12, 1966	Flight 32

Name: Gemini 11 (GTA-11)
Sequence: 32nd astro-flight, 23rd spaceflight, 21st Earth orbit
Launch date: September 12, 1966
Launch site: Pad 19, Cape Kennedy, USA
Launch vehicle: Titan II
Flight type: Earth orbit, rendezvous, docking, re-boost and spacewalk
Flight time: 2 days 23hr 17min 8sec
Spacecraft weight: 8,374lb
Crew: Cdr Charles "Pete" Conrad Jr, 36, USN, command pilot
Lt-Cdr Richard Francis Gordon Jr, 37, USN, pilot

Like each flight before it in the programme, Gemini 11 was designed to do that something extra, capitalising on the success of previous flights and practising all the manoeuvres needed on the road to the Moon. For command pilot Charles Conrad and pilot Richard Gordon the challenge included a launch within a 2sec window; a docking with Agena 11 on the first orbit in what was called a direct-ascent rendezvous, simulating an emergency lunar take-off or abort; a prolonged spacewalk by Gordon in which, it was hoped, the EVA problems encountered by his predecessors would be ironed out; a record 850-mile push into space; scientific experiments; and the first attempt to create artificial gravity in space.

As Gemini 8 back-up crew the two astronauts had had ample training in rendezvous and docking, and their expertise showed when, 94min after launch at 9.42 am on September 12, 1966, Gemini 11 docked with Agena 11 over the Pacific Ocean. For the first time on a rendezvous mission

Pad 19, Cape Kennedy, September 12, 1966. Conrad leads Gordon to the spacecraft (Nasa)

A classic photo of Earth from space, taken from a height of 540 miles from Gemini 11 during its high-altitude flight on the power of the docked Agena target vehicle. India and Sri Lanka can be seen clearly (Nasa)

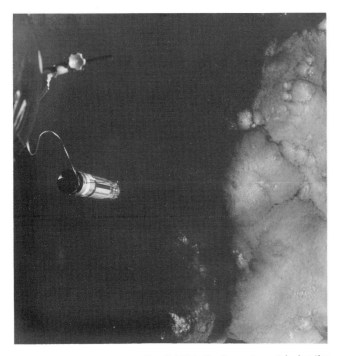

A 100ft tether line connects Gemini 11 to the Agena target during the first creation of artificial gravity in space (Nasa via Astro Information Service)

the pilot, Gordon, had a few tries at docking.

Born in Seattle, Washington, on October 5, 1929, Richard Gordon graduated from the University of Washington with a science degree and entered flight training in 1951. In 1961 he emulated John Glenn by breaking the trans-US speed record by flying from Los Angeles to New York, in 2hr 47min, an average speed of 869mph, to win the Bendix Trophy. Before being selected as a Group 3 astronaut in October 1963 Gordon had attended the All-Weather Flight School, the US Naval Postgraduate School and the Test Pilot School.

The next experiment on the schedule after the first-orbit docking was Gordon's planned 155min EVA. One of his main jobs as he floated outside the capsule was to attach a tether to the Agena target in preparation for the artificial gravity experiment. On the ground during training this had taken about 30sec, but in space Gordon found he was continually fighting against the tendency to float upwards. This excess work created too much heat for the temperature-control system of his spacesuit, and his visor became fogged over with sweat. "I'm pooped, Pete," he said, ". . . I still can't see. The sweat won't evaporate." After 38min in space, during which he did eventually manage to tie the tether, he was recalled by the command pilot. Conrad explained the situation to ground control: "Listen, I just brought Dick back in. He got so hot and sweaty he couldn't see." Yet again a spacewalk had been cut short. At the time it seemed as though EVA might be the one insurmountable hurdle on the way to the Moon. Could the astronauts work effectively outside the spacecraft? The answer would have to wait until the final Gemini flight.

The spacewalk over, the astronauts commanded the Agena main engine to re-start and send them into a higher orbit. The motor fired first time and the Agena/Gemini combination soared up to a maximum height of 850 miles, passing over the Middle East at about 250 miles altitude, the Indian Ocean, India and Sri Lanka at about 500 miles, and North-western Australia at 850 miles. The view staggered the two astronauts, and as the craft came up over Australia Conrad exclaimed: "The world is round!" For the first time ever an astronaut had seen the Earth as a globe. He continued: "It's utterly fantastic. You wouldn't believe it. It really is blue. The water really stands out and everything is blue. The curvature shows up a lot. Looking straight down you still see just as clearly. There is no loss of colour and the detail is extremely good." The orbit was then lowered in preparation for Gordon's second space excursion, this time standing on his seat for 128min to take photographs of the stars, the Earth and other objects.

Perhaps one of the most interesting features of the Gemini 11 mission was the creation of artificial gravity in an experiment designed to test a method of achieving passive stabilisation in orbit. With the 100ft tether joining the two craft together, Conrad separated the Agena and Gemini until the tether went taut. He then fired the thrusters, sending the two craft into a slow spin. The resulting reaction registered on the cabin instruments as a non-weightless condition. The combination flew like this for two orbits before the Agena was discarded and the astronauts got down to ten more scientific tasks and another rendezvous exercise.

For the landing in the Atlantic Ocean 700 miles east of Miami the astronauts had something new in store. They just sat back and folded their arms while the spacecraft's computer worked out the alignment for re-entry, fired the retro-rockets and flew them back to Earth to within 1½ miles of the recovery ship, USS *Guam*. They were on deck in 23min, a record for the Gemini programme. After their 71hr 17min, 1,232,530-mile, 44-revolution flight Conrad said: "This old world looks good from the deck of the carrier. But boy it looks really great from 850 miles out. We had a very good flight. A good night's sleep and we'll be ready to go up again."

Name: X-15 No 3, Flight No 56
Sequence: 33rd astro-flight
Flight date: November 1, 1966
Take-off site: Edwards Air Force Base, California, USA
Launch vehicle: B-52 carrier aircraft
Flight type: Sub-orbital
Flight time: 10min 43sec
Spacecraft weight: About 34,000lb
Crew: William Harvey Dana, 35

William Dana was born in Pasadena on November 3, 1930, and received a science degree from the US Military Academy in 1952 before joining the USAF. He left the service to study aeronautical engineering, graduating in 1958 and then joining Nasa as a research pilot. Dana was chosen in 1965 as the 11th X-15 pilot and made his first flight on the day after his birthday that year. In 1966 Dana flew five missions, the last of which qualified him as an astronaut, albeit unofficially.

Tasks for the flight included the collection of micro-meteorites, tests of a dual-channel radiometer and a tip-pod accelerometer, and precise measurements of altitude and atmospheric density. It was the 174th X-15 mission and the 56th by X-15 No 3. Maximum speed was 3,750mph (Mach 5.46) and peak altitude 306,900ft (58 miles).

Bill Dana, the 35th man above 50 miles (Nasa)

Gemini 12   November 11, 1966   Flight 34

Name: Gemini 12 (GTA-12)
Sequence: 34th astro-flight, 24th spaceflight, 22nd Earth orbit
Launch date: November 11, 1966
Launch site: Pad 19, Cape Kennedy, USA
Launch vehicle: Titan II
Flight type: Earth orbit, rendezvous, docking and spacewalk
Flight time: 3 days 22hr 34min 31sec
Spacecraft weight: 8,297lb
Crew: Cdr James Arthur Lovell Jr, 38, USN, command pilot
Maj Edwin Eugene "Buzz" Aldrin Jr, 36, USAF, pilot

Buzz Aldrin (left) and James Lovell acknowledge the cheers of sailors aboard the recovery ship after their flight completed the Gemini programme (Nasa)

When Gemini 12 splashed down in the Atlantic Ocean after a four-day flight America had chalked up almost 2,000 man-hours in space. The way to the Moon had been cleared, the Soviet Union was far behind, and it seemed incredible that in the short space of just a year and a half the whole picture could have changed so completely.

Gemini 12 was not without its problems, but it ended the programme with another success, the twelfth in twelve attempts, ten with men aboard. The command pilot was James Lovell of Gemini 7, who also served as Gemini 9 back-up command pilot, and the pilot was Edwin "Buzz" Aldrin. The astronauts walked to the launch pad with cards on their backs reading "The End". When they got to the spacecraft they saw a sign fixed to it: "Last chance. No reruns. Show will close after this performance."

Aldrin was nicknamed "Buzz" as a child in the Monclair, New Jersey, home where he was born on January 20, 1930. The name was to stick with him right up to his astronaut days. Like most of the 1963 class of astronauts he was a holder of a science degree, but he was also a veteran of the Korean War, having flown 66 combat missions in F-86s for the USAF. He subsequently served with a number of other squadrons in the US and Europe, and before his selection as an astronaut he had earned a doctorate in astronautics at MIT, where he wrote a thesis on manned orbital rendezvous. He was to put the theories into practice on November 11, 1966, when the final Gemini-Titan combination left Pad 19 at 3.46 pm with the precision and matter-of-factness of a mainline train from a city station.

About 90min earlier, the Agena 12 target had been inserted into orbit. The link-up proved more difficult than anticipated because the astronauts lost their radar lock on the target and had to rendezvous visually, with Lovell piloting the craft under directions from the quick-minded Aldrin, who worked out the necessary data using a sextant. Rendezvous was achieved at T+3hr 46min and docking at T+4hr 14min. Suspecting a potentially dangerous malfunction, technicians on the ground decided to stop Lovell and Aldrin from firing the Agena engine to take them up to 400 miles. Instead the astronauts manoeuvred into an orbit which would allow them, as planned, to see a total eclipse from space for the first time ever. After photographing the eclipse over South America on orbit 10 Aldrin made the first of his three EVAs. Standing on his seat, he spent 2hr 39min taking photographs and making various observations.

On the second day minor snags hit Gemini 12 when some of the manoeuvring thrusters began to misbehave slightly. The problem was not severe, however, and Aldrin made the spacewalk which represented Nasa's last chance to get over the EVA problems encountered by the earlier astronauts. Aldrin had harnesses with which to attach himself to the two spacecraft, and Velcro pads on his gloves which stuck to

areas of the same material on the craft so that he could manoeuvre himself without having to fight continuously against the tendency to float upwards. A pair of "golden slipper" foot restraints had also been installed at the work station on the Gemini, allowing Aldrin to perform his tasks without struggling to stay in one place. Aldrin's highly successful EVA lasted 2hr 8min, during which time his only complaint was of slightly cold feet.

One of Aldrin's other jobs had been to join the two craft by a tether in preparation for a repeat of the artificial-gravity experiment first carried out by Gemini 11. Aldrin made his third EVA on the third day, standing on his seat for 51min and becoming the EVA record-holder with a total of 5hr 30min. The crew also conducted 14 science experiments.

The final day of the Gemini programme saw the crew beset with minor problems. The thruster problem worsened, the astronauts ran out of water because of malfunctioning fuel cells and had to use a back-up supply, and there were control problems just before re-entry. But Gemini 12 made it back, landing in the Atlantic 2.6 miles from USS *Wasp* after 94hr 34min, 59 revolutions and 1,628,510 miles. "A four-day vacation with pay and you see the world," said Lovell.

Gemini had cleared the way to the Moon at a cost of $1,283,400,000 and had clocked up over 80 man-days in space. Apollo beckoned and attention moved up Cape Kennedy's "Gantry Row" to Pad 34.

Agena target vehicle is launched from Pad 14 at 2.08 pm, 90min before Gemini 12 (Nasa)

Sightseer in space. Aldrin casually leans out of the Gemini 12 hatch during a stand-up EVA (Nasa via Astro Information Service)

Soyuz 1 April 23, 1967 Flight 35

Name: Soyuz 1
Sequence: 35th astro-flight, 25th spaceflight, 23rd Earth orbit
Launch date: April 23, 1967
Launch site: Tyuratam, USSR
Launch vehicle: A2 (SL-4)
Flight type: Earth orbit
Flight time: 1 day 2hr 47min 52sec
Spacecraft weight: 14,222lb
Crew: Col Vladimir Mikhailovich Komarov, 40, Soviet Air Force

The year 1967 was a tragic one for space exploration. Three Americans died in Apollo 1 during a ground test on the launch pad, and the Soviet Union's second-generation spacecraft made a disastrous debut.

Soyuz, Russia's first truly new manned craft since the pioneering Vostok, was 26ft long, weighed 14,000lb, consisted of three modules and derived its electrical power from two solar panels. At the front of the craft was a docking probe or collar to facilitate the joining of two craft together; Soyuz, after all, is the Russian word for "union." The docking mechanism was attached to a spherical orbital module which acted as an airlock on joint flights or as a laboratory on solo missions. Attached to the orbital module was the flight cabin, shaped like an inverted cup and with a heatshield at the base. This was equipped with a single main parachute and an emergency back-up. The opening of the main chute was to be preceded by drogue deployment at 25,000ft. Just before landing, at about 3ft above the ground, deceleration rockets would fire to reduce the impact speed to about 1ft/sec. Behind the flight capsule was an instrument section to which were attached the two wing-like solar panels, each 12ft long. The instrument section was equipped with two liquid-propellant rocket motors, each with a thrust of 880lb; one of these units was a back-up. When stacked on top of the standard A2 launcher the whole assembly was covered with a payload shroud surmounted by a three-tier rocket escape tower designed to pull the craft away in the event of a booster malfunction.

Veteran cosmonaut Vladimir Komarov, callsign *Ruby* and the first Russian to make two spaceflights, boarded Soyuz 1 early on April 23, 1967, and was blasted into space at 3.35 am

Vladimir Komarov, the lone pilot of the new Soyuz 1 spacecraft. He is pictured in a spacesuit similar to those worn by the Voskhod 2 crew. On his first flight (Voskhod 1) he did not have a spacesuit, and nor did the crews on a number of subsequent Soyuz flights (Novosti)

Komarov's widow at the Kremlin Wall, in which the cosmonaut's ashes had been placed. On the right is Konstantin Feoktistov (Novosti)

Moscow time. It was the first manned launch at night, and the plan was for Soyuz 1 to dock with Soyuz 2, to be launched the following day with Valeri Bykovsky, Yevgeni Khrunov and Alexei Yeliseyev aboard. The official statement used the words "extended flight". But Soyuz 1 – in a 125/139-mile orbit at an inclination of 51°, as opposed to the 65° of earlier Russian missions – proved to be a catastrophe.

Unofficial reports indicate that although one solar panel had failed to deploy the mission was going tolerably well until the 13th orbit. Then the stabilisation system failed, causing Soyuz to go into a debilitating Gemini 8-like spin. As if this was not bad enough, there had been failures of the on-board computer, an antenna and a TV transmitter. Komarov attempted to re-enter on the standard 16th orbit but failed. He tried again on the 17th orbit, once more without success. In a last-ditch effort the retros were fired on orbit 18 after, as reports from a US listening post in Turkey indicated, Komarov had accepted that he was doomed and had spoken to his wife and to Premier Kosygin, who assured him that he would always be remembered. It would appear that the flight module was out of control when it separated from the orbital section and the instrument module, resulting in excessive re-entry heat loads on essential systems, including the vital parachute.

According to the Soviet account, the main parachute tangled and failed when an attempt was made to deploy it at 25,000ft. Soyuz 1 plummeted like a stone and smashed into the ground. Komarov, who mercifully might already have been dead, could not have abandoned the capsule, which was not fitted with an ejection seat. After a long silence the Russians announced the tragic news, mentioning only the parachute failure.

Soyuz 1 had lasted 26hr 40min and the ill-fated craft had travelled about 480,000 miles. Komarov's ashes were placed in the Kremlin Wall during a state funeral watched by millions on television. He had been the first man to die during a space mission. The Soviet comeback in space had ended catastrophically and it would be a year and a half before the Russians tried again. The consequences of the Apollo 1 fire the previous January were similar, resulting in an 18-month hiatus in the world's manned orbital spaceflight effort.

X-15 No 3, Flight No 64   October 17, 1967   Flight 36

Name: X-15 No 3, Flight No 64
Sequence: 36th astro-flight
Flight date: October 17, 1967
Take-off site: Edwards Air Force Base, California, USA
Launch vehicle: B-52 carrier aircraft
Flight type: Sub-orbital
Flight time: 10min 6sec
Spacecraft weight: About 34,000lb
Crew: Maj William "Pete" Knight, 38, USAF

Pete Knight, the 37th man above 50 miles (Nasa)

For Maj "Pete" Knight 1967 proved to be a special year. He qualified as an astronaut and became the fastest man in the atmosphere.

Born in Noblesville, Indiana, on November 18, 1929, Knight joined the USAF in 1951 and graduated from the USAF Institute of Technology with a degree in mathematics. He served as a fighter pilot and maintenance officer before entering the USAF Experimental Test Pilot School. In March 1962 he was selected to join the group of five USAF and one civilian astronauts nominated to fly the Dyna-Soar winged space glider. The project was cancelled in late 1963, and in 1965 Knight joined the X-15 programme. He made his first flight on September 30 of that year and was later named as pilot of the converted X-15 No 2, designed for ultra-high-speed research. The first of two world record-breaking flights came on November 18, 1966, when Knight flew his eighth X-15 mission and the 175th of the programme. At 98,000ft maximum altitude was comparatively low because all the rocket power was used to generate speed rather than propel the X-15 upwards. Knight reached the highest point within seconds, then nosed down and power-dived to a speed of 4,250mph (Mach 6.33). On his 12th and the programme's

188th mission on October 3, 1967, the speed record was pushed up to 4,520mph (Mach 6.7).

On his next flight, in X-15 No 3 on October 17, Knight was again in the news, this time as an unofficial astronaut. His tasks for the mission were to collect micrometeorites, record wing-tip pod deflection during re-entry, observe the ultra-violet plume of the rocket exhaust, and to study the solar spectrum above 200,000ft. Knight reached an altitude of 280,500ft (53.4 miles) and a maximum speed of 3,856mph (Mach 5.33). It was the 64th mission of X-15 No 3. The aircraft's next flight also took it above 50 miles, but with tragic results.

X-15 air drop and ignition (Nasa via Astro Information Service)

X-15 No 3, Flight No 65   November 15, 1967   Flight 37

Name: X-15 No 3, Flight No 65
Sequence: 37th astro-flight
Flight date: November 15, 1967
Take-off site: Edwards Air Force Base, California, USA
Launch vehicle: B-52 carrier aircraft
Flight type: Sub-orbital
Flight time: 4min 51sec
Spacecraft weight: About 34,000lb
Crew: Maj Michael James Adams, 37, USAF

Long before Michael J. Adams made his seventh X-15 flight and the 191st of the programme, the whole business of rocketing into near-space had become routine and apparently without hazard. Adams' final mission was to be a sharp reminder of the dangers of space exploration.

Mike Adams was born in Sacramento on May 5, 1930, joined the Air Force on November 22, 1952, and flew 49 missions in the Korean War. He later graduated in aeronautical engineering and astronautics before joining the Aerospace Research Pilot School at Edwards AFB. After completing his training there he was one of four pilots to take part in a five-month series of Nasa lunar landing trials in 1964. A year later Adams became one of the first military astronauts, joining the group of pilots chosen for the Manned Orbital Laboratory programme. Then, frustrated by delays in MOL (which was ultimately cancelled in 1969), Adams replaced Joe Engle when he left the X-15 programme in 1966.

On his first X-15 mission, on October 6, 1966, Adams performed admirably when the engine failed to light up and he had to glide the aircraft back to a premature landing. He made five more flights before the fatal mission of November 15, 1967. Tasks for this flight were an ultra-violet study of the rocket exhaust plume, observations of the solar spectrum and the bow shockwave of the wing-tip pod, monitoring of nose-gear loads, collection of micrometeorites, and tests of the ablative material used on the Saturn V booster.

X-15 flight No 191, and the 65th by X-15 No 3, reached 3,570mph (Mach 5.20) and a maximum altitude of 266,000ft (50.4 miles). Adams had thus become an astronaut, but suddenly he was in trouble. At 3,000mph and a height of 230,000ft the X-15 entered a spin. The aircraft suddenly yawed, and in trying to correct Adams misread his instruments and yawed the X-15 still further. "I'm in a spin,

Michael Adams, the 38th man above 50 miles (USAF)

Pete!" he reported to Pete Knight, the flight communicator. As the X-15 descended the spin continued, subjecting the aircraft and pilot to tremendous forces. At 120,000ft Adams recovered from the spin but then the vehicle developed a pitch oscillation that was self-sustaining and increasing in severity. "I'm in a spin ... I'm in a spin," Adams called desperately. The X-15 was like a leaf in a storm.

The rocket aircraft tumbled and twisted, descending at a speed of 2,600ft/sec. Adams, who was enduring alternating 15g negative and positive forces, was powerless. "Let's keep it up, Mike, let's keep it up," said Pete Knight. But it was too late. At 60,000ft the X-15 broke apart. The fuselage, wings and three control surfaces fell separately, strewing wreckage over a 10 × 1.5-mile area three miles north-east of Johannesburg, California. A spotter plane radioed to Knight: "I've got a lot of dust on the lake down there." In the mangled cockpit Michael Adams was found dead, still strapped to his seat.

Name: X-15 No 1, Flight No 79
Sequence: 38th astro-flight
Flight date: August 21, 1968
Take-off site: Edwards Air Force Base, California, USA
Launch vehicle: B-52 carrier aircraft
Flight type: Sub-orbital
Flight time: 9min 23sec
Spacecraft weight: About 34,000lb
Crew: William Harvey Dana, 37

Bill Dana pictured after an X-15 mission on March 1, 1968, when he reached an altitude of 105,000ft (Nasa)

The 13th and final X-15 astro-flight was completed on August 21, 1968, during the 197th mission in the programme. Flying X-15 No 1 on its 79th mission and his own 15th, Bill Dana reached 3,443mph (Mach 5.01) and a height of 267,500ft (50.7 miles). It was Dana's second X-15 astro-flight.

The programme was due to end on the 200th flight, so as Bill Dana came in to land on October 28, 1968, at the end of the 199th flight he was unaware that there would in fact be no more X-15 missions. The 200th was attempted twice but cancelled due to bad weather before it was decided to scrap the flight altogether. Flight time for the 199 missions was 30hr 13min 49.4sec and a total of 41,763.8 miles had been travelled. Experience at Mach 4 and 5 totalled nearly 6hr and 1½hr respectively, plus 1min 18sec at Mach 6. X-15 No 1 is now at the National Air and Space Museum in Washington DC and No 2 is displayed at the USAF Museum in Ohio.

Apollo 7   October 11, 1968   Flight 39

Name: Apollo 7 (AS-205)
Sequence: 39th astro-flight, 26th spaceflight, 24th Earth orbit
Launch date: October 11, 1968
Launch site: Pad 34, Cape Kennedy, USA
Launch vehicle: Saturn IB (SA-205)
Flight type: Earth orbit
Flight time: 10 days 20hr 9min 3sec
Spacecraft weight: 32,395lb
Crew: Capt Walter Marty Schirra Jr, 45, USN, commander
Maj Donn Fulton Eisele, 38, USAF, senior pilot
Walter Cunningham, 36, pilot

On January 27, 1967, astronauts Wally Schirra, Donn Eisele and Walter Cunningham were serving as the Apollo 1 back-up crew, with a view to making a Moon flight later. Then, with horrifying suddenness, they found themselves promoted to replace a crew who now lay dead in their burned-out spacecraft. Virgil Grissom, Edward White and Roger Chaffee had been carrying out a ground run-through inside their command module atop an unfuelled Saturn IB launcher at Pad 34 when a fire, caused by a short circuit and nourished by a pure oxygen atmosphere, exploded into an inferno within seconds. The disaster delayed the programme for nearly two years and some people doubted that America would get to the Moon before the end of the 1960s.

Apollo consisted of three components, the command, service and lunar modules. The first Apollo flights were intended to test the command and service modules only. The command module (CM) was a cone-shaped capsule 12ft high and 12ft 10in in diameter at the base, and weighing 12,250lb. It was equipped with heatshields, 12 reaction-control motors, landing systems, computers and a docking probe and transfer hatch for operations with the lunar module (LM). The service module (SM), 12ft 10in in diameter and 24ft 7in long, weighed 51,240lb and incorporated fuel cells, the service propulsion system (SPS) with a thrust of 21,500lb, reaction-control motors and communications antennae. Apollo 7 was a simpler version of the basic spacecraft, weighing only 30,000lb, and did not include the full suite of communications equipment. At launch the crew breathed an oxygen-nitrogen atmosphere.

The first test flight of a manned CM/SM was to be launched by a Saturn IB. Topped by an Apollo with emergency escape tower, the combination was 224ft tall. The Saturn IB first stage consisted of eight H-1 engines developing a total thrust of 1,640,000lb. The second stage, the S-IVB, was to be the third stage of the Apollo-Saturn V combination used for Moon flights. This was powered by a restartable J-2 engine fuelled with liquid oxygen and liquid hydrogen and developing 225,000lb of thrust. Instead of Gemini-style ejection seats Apollo was equipped with a launch escape system (LES) comprising a tower-mounted, 33,000lb-thrust solid rocket motor attached to the apex of the boost protective cover which shrouded the command module for the launch. The LES was jettisoned once the spacecraft had reached a point

Apollo 7 crew framed by the spacecraft's quick-release hatch: (left to right) Donn Eisele, Wally Schirra and Walter Cunningham (Nasa via Astro Information Service)

beyond which the main recovery systems could be used.

In between the first and second stages of the Saturn IB was the instrument unit (IU), housing systems for the guidance, radio communications, telemetry, environmental control and electrical power of the rocket-spacecraft combination.

Apollo 7's task was to prove the Apollo system – in particular the major redesign resulting from the Apollo 1 fire – and check out the worldwide Manned Spacecraft Tracking Network. Schirra and his crew worked harder and closer to their spacecraft than any before them, such was their near-obsession with making a success of a flight that was the key to a Moon landing in 1970.

Schirra, now 45 years old and making his third spaceflight, was nicknamed "Mr Cool" by the space agency staff. His crew was a fresh one, consisting of two Group 3 rookies. The senior pilot was Maj Donn F. Eisele, born in Columbus, Ohio, on June 23, 1930. He joined the US Naval Academy and received a science degree before entering the Air Force. He was a graduate of the USAF Aerospace Pilot School and also held a degree in astronautics.

The pilot was Walter Cunningham, a civilian scientist-pilot. Born on March 16, 1932, in Creston, Iowa, he joined the Navy in 1951, earning his pilot's wings in 1952. He then concentrated on science and by the time of his selection as an astronaut in October 1963 he had earned two physics degrees and one doctorate and had worked as a research scientist for the Rand Corporation.

Apollo 7 was to be a ten-day engineering shakedown in which every system, switch and wire would be tested thoroughly to ensure the Moonworthiness of the spacecraft. At 11.00 am on October 11, 1968, the Saturn IB blasted away from Cape Kennedy's Pad 34, the scene 21 months earlier of the tragedy that had created a near-fatal crisis of confidence within the Apollo programme. The flight that followed was to more than make up the lost ground.

"We're having a ball . . . she's riding like a dream," said Schirra as Apollo 7 made its first orbit around the Earth. The first manoeuvre was the separation of the command and

Launch of Apollo 7 from Pad 34 at Cape Kennedy (Nasa via Astro Information Service)

service modules from the S-IVB second stage, followed by a 180° rotation and approach to the stage to simulate the extraction of a lunar module encased in the S-IVB. The astronauts manoeuvred to within 70ft of the stage, station-keeping for 20min and taking superb photographs of the S-IVB right over the Cape.

Fifteen hours into the flight, the mood had changed. Schirra reported that he had a bad cold, had used up a number of handkerchiefs and had taken two asprins. His illness probably had something to do with his occasional testiness during the flight, first noticed when he refused to turn on a 4½lb black-and-white camera for the first in-flight television show of the mission. The urgings of ground control prompted this tirade from Schirra: "You have added two burns to this flight schedule, you have added a urine water dump and we have a new vehicle up here, and I tell you this flight TV will be delayed without further discussion until after the rendezvous . . . we don't have the equipment, we have not had an opportunity to follow setting, we have not eaten at this point, I still have a cold. I *refuse* to foul up our timelines this way!"

The rendezvous he referred to was one of many exercises – involving SPS burns and subsequent manoeuvres – designed to simulate the en route activities of an actual lunar mission. The flight schedule included eight burns of the SPS, the longest of which lasted 66sec and significantly changed the spacecraft's orbit. The highest point reached during the flight was 269 miles. When the engine fired for the first burn Schirra, more lighthearted this time, said: "That was a real kick in the pants!" He also finally relented on the television shows, which were spectacular, good-humoured and aimed particularly at younger viewers. "Hello from the lovely Apollo room, high atop everything," was the introduction to the shows.

The fun did not last long, however, and the flight continued with further altercations between space and ground about the additional experiments requested by mission control. At one point Schirra shouted: "I've had it up to here today and from now on I'm going to be an on-board flight director." The doughty senior astronaut continued to assert himself right to the end of the mission. Because they still had colds Schirra wanted the crew to re-enter with their helmets off. The ground ordered them to put their helmets on. Schirra insisted and they came home without them.

After an 11sec retrofire and re-entry over Texas at 400,000ft Apollo 7 appeared under its three parachutes at 27,000ft and the capsule splashed into the Atlantic about eight miles from USS *Essex*. At first the command module floated upside down in Stable 2 position, but buoyancy balloons soon righted the craft and it was spotted by recovery crews 22min later.

The 11-day, 163-revolution, 4.5 million-mile flight, the second longest to that date, had been a brilliant success. It had thoroughly vindicated the Apollo system and raised hopes of a landing on the Moon the following year.

Soyuz 3   October 26, 1968   Flight 40

Name: Soyuz 3
Sequence: 40th astro-flight, 27th spaceflight, 25th Earth orbit
Launch date: October 26, 1968
Launch site: Tyuratam, USSR
Launch vehicle: A2 (SL-4)
Flight type: Earth orbit and rendezvous
Flight time: 3 days 22hr 50min 45sec
Spacecraft weight: 14,500lb
Crew: Col Georgi Timofeyevich Beregovoi, 47, Soviet Air Force

After achieving the automatic docking of two Soyuz craft under the guise of Cosmos 186 and 188 in October 1967 the Russians were ready to try the exercise with manned vehicles. They had also looped the Moon with a Soyuz-based Zond spacecraft and safely returned it to Earth, leading to rumours that one man would fly around the Moon before the end of 1968. So, when the unmanned Soyuz 2 was launched on October 25, 1968, and the oldest man in space, Georgi Beregovoi, followed in Soyuz 3 a day later, the world's newsrooms were buzzing with excitement. Would Beregovoi do a lunar loop? As it happened, he did not even join up with Soyuz 2.

Beregovoi was born on April 15, 1921, in Fyodorovka in the Poltava region of the Ukraine and joined the Soviet Air Force at the age of 17. He flew 185 combat missions during the Second World War and was made a Hero of the Soviet Union. Beregovoi then entered test pilot training and in 1961 was presented with the Merited Test Pilot of the USSR award. He joined the cosmonauts in 1964.

With Soyuz 2 already aloft in its 51° orbit on October 26, Beregovoi, callsign *Argon*, rose to meet it. His launch was televised very soon afterwards and for the first time Western viewers caught a glimpse of the A2 rocket and of a Soviet lift-off the day it happened. An automatic rendezvous procedure took Beregovoi to within 656ft of Soyuz 2. Soyuz 3 moved closer under manual control but then backed off to about a mile, went into another orbit and performed a second rendezvous. It is thought that Beregovoi experienced control shortcomings during the critical close-in manoeuvring.

On October 28 the re-entry module of Soyuz 2 made an automatic return to Earth and Beregovoi continued his flight, spending the time showing television viewers around his spacecraft, making observations of the Earth and stars and performing further manoeuvres. Soyuz 3 had completed 64 orbits when its retro-rocket fired for 145sec to slow it down for re-entry. Beregovoi jettisoned the orbital module and the instrument section and made a safe re-entry, landing at T+3 days 22hr 51min in Karaganda to become the second most travelled cosmonaut to that date.

Soyuz 3 pilot Georgi Beregovoi. Soyuz 3 was the 25th manned orbital flight (Novosti)

Name: Apollo 8 (AS-503)
Sequence: 41st astro-flight, 28th spaceflight, 26th Earth orbit, 1st lunar flight, 1st lunar orbit
Launch date: December 21, 1968
Launch site: Pad 39A, Kennedy Space Centre, Merritt Island, USA
Launch vehicle: Saturn V (SA-503)
Flight type: Lunar orbit
Flight time: 6 days 3hr 0min 42sec
Spacecraft weight: 63,717lb
Crew: Col Frank Borman, 40, USAF, commander
Capt James Arthur Lovell Jr, 40, USN, senior pilot
Maj William Alison Anders, 35, USAF, pilot

Christmas 1968 will be remembered as the time men first went to the Moon. Three factors lay behind this sudden leap across Earth-Moon space: delays in lunar module development, so that Apollo 8, due to fly a lunar module in Earth orbit, was pushed back and Apollo 9 brought forward to fly (as Apollo 8) without a lunar module but in a high Earth orbit; the high confidence engendered by Apollo 7's success; and, perhaps most important, the Soviet Union's Zond unmanned circumlunar flight. With all this in mind, on November 12, 1968, Nasa chief Dr Thomas Paine gave the go-ahead for an Apollo 8 Moon mission.

Veterans Frank Borman and James Lovell, the latter replacing an injured Michael Collins, were teamed for Apollo 8 with rookie William Anders. But, as Anders himself said, they were all rookies when it came to flying the huge Saturn V, which was being entrusted with its first human payload on only its third flight. Standing 363ft high, it comprised three stages. The first, designated S-IC, had five F-1 engines developing a total thrust of 7,590,000lb. The S-II second stage was powered by five J-2 engines giving a total of 1,000,000lb, and the third stage was the S-IVB. The whole Apollo-Saturn assembly weighed about 6,400,000lb. Later in the programme the thrust of the S-IC was increased to 7,723,000lb, that of the S-II to 1,150,000lb and that of the S-IVB to 238,000lb as ratings varied from mission to mission.

Anders was born in Hong Kong on October 17, 1933. He earned a science degree at the US Military Academy before entering the USAF and subsequently acquiring a degree in nuclear engineering. After his selection in Nasa's third astronaut group he served as Gemini 11 back-up pilot. Trained as lunar module pilot for Apollo 9, he must have been galled to find himself flying without one.

At 7.51 am on December 21 at Pad 39A on Merritt Island, just north of Cape Kennedy, the epic mission began with first-stage ignition at T-8sec. The mammoth Saturn seemed to hover, belching flame and steam, before rising ponderously, gulping 15 tons of fuel a second. At T+11min 25sec Borman reported third-stage cut-off, and Apollo 8 was in Earth parking orbit. The S-IVB re-ignited at T+2hr 50min and with a 5min 17sec burn propelled Apollo 8 away from Earth at a speed of 24,226mph. Soon after the S-IVB had been jettisoned the Earth came under the enthusiastic scrutiny of the astronauts. Lovell described what he could see out of one of the seven windows of the craft: "I can clearly see the terminator, I can see most of South America all the way up to

Central America, Yucatan and the peninsula of Florida." Anders then chimed in: "Tell the people in Tierra del Fuego to put on their raincoats. Looks like a storm out there."

The flight towards the Moon was punctuated by two mid-course corrections, achieved with the mission's first firings of the SPS engine, and television transmissions. "I certainly wish that we could show you the Earth," said Borman (the light was too strong for the on-board television camera). "It is a beautiful, beautiful view with a predominantly blue background and just huge covers of white clouds . . . We are all in very good shape . . . It was a very exciting ride on that Saturn, but it worked perfectly."

The craft was steadily slowing as it approached the point at which the Moon's gravitational influence exceeded that of the Earth. Apollo 8's speed had fallen to a mere 2,223mph at a distance of 38,900 miles from Earth's satellite, but then lunar gravity took charge and the spacecraft began accelerating "downhill" towards its objective. This important milestone prompted some musings from Lovell: "What I keep imagining is, if I were a traveller from another planet, what would I think about the Earth from this altitude . . . whether I think it would be inhabited . . . I was just kind of curious if I would land on the blue or the brown part." Borman replied: "You better hope we land on the blue part."

A little later the capsule communicator at Houston, astronaut Gerald Carr, reminded the crew of their position in

Apollo 8's Moon-bound crew pose outside the VAB for the rollout of their Saturn V booster. Left to right: Frank Borman, James Lovell and William Anders (Nasa via Astro Information Service)

Just after the lunar orbit insertion burn Apollo 8 came over the horizon and saw Earthrise. Earth, 240,000 miles away, is 5° above the lunar horizon, which is about 430 miles from the spacecraft. The picture spans about 100 miles of lunar surface (Nasa)

space: "By the way, welcome to the Moon's sphere."

"The Moon's fair?" replied a puzzled Anders.

"The Moon's sphere – you're in the influence," said Carr.

"That's better than being under the influence, hey Gerry?," joked Anders.

On December 24 Apollo 8 swung behind the Moon at T+68hr 58min 45sec and lost contact with the Earth. The SPS fired for 242sec, slowing the craft from 5,000mph to a lunar orbital speed of 3,721mph. At first only the astronauts knew if the burn had been successful, with the bulk of the Moon still cutting off Earth-spacecraft communications. If the engine had failed to fire the craft would have swung around the Moon and headed back to Earth on a free-return trajectory. But some 35min after the burn contact was re-established as the craft came up over the eastern horizon of the Moon.

The first description of what the Moon looked like in close-up came from James Lovell: "The Moon is essentially grey, no colour. Looks like plaster of paris. Sort of greyish sand." During the television programmes beamed to an excited Earth audience, all three lunarnauts, as they had been

dubbed, gave their own descriptions:

Borman: "The Moon is very bright and not too distinct in this area."

Anders: "The colour of the Moon looks like whitish grey, like dirty beach sand with lots of footprints on it. Some of these craters look like pickaxes striking concrete, creating a lot of fine haze dust."

Lovell: "As a matter of interest, there's a lot of what appears to be small new craters that have these little white rays radiating from them. There is no trouble picking out features that we learned on the map."

On the third orbit the path around the Moon was circularised at 70 miles altitude and the heavy schedule of observational and photographic experiments began. Lovell and Anders were tired at this point and Borman decided to cancel some of the experiments to allow his crew to get some well earned rest. "Lovell is snoring already," said Borman a few minutes later. The capsule communicator replied: "Yeah, we can hear him down here."

During the ninth orbit, on Christmas Day, the second

television transmission was beamed to Earth and Borman, Lovell and Anders gave their final impressions of the Moon:

Borman: "Its a vast, lonely, forbidding type of existence, a great expanse of nothing . . . It would not appear to be a very inviting place to live and work."

Lovell: ". . . the vast loneliness of the Moon up here is awe-inspiring and it makes you realise just what you have back there on Earth. The Earth from here is a grand oasis to the vastness of space."

Anders: "I think the thing that has impressed me most is the lunar sunrises and sunsets."

At the end of the transmission Anders said: "For all the people back on Earth, the crew of Apollo 8 has a message that we would like to send to you." Then he, Lovell and the commander, in that order, read out passages from the Book of Genesis: "In the beginning God created the Heaven and the Earth . . ." Borman then said goodbye with: "And from the crew of Apollo 8, we pause with goodnight, good luck, a merry Christmas and God bless all of you – all of you on the good Earth."

After one further orbit of the Moon came the penultimate drama of the mission. The SPS engine was to fire once more to send Apollo 8 out of lunar orbit, in which it had travelled for 20hr 11min, and on to its course for Earth. The most tense moments for the people on the "good Earth" came while Apollo 8 was behind the Moon. But the 203.7sec burn was successful and as the spacecraft came round the Moon the reassuring voice of Jim Lovell could be heard: "Please be informed there is a Santa Claus."

At the end of the trans-Earth flight the command module parted company with the service module and seared into the atmosphere at a speed of 24,696mph. The angle of entry into the atmosphere was particularly critical. If the capsule had skimmed into it at a shallow angle it would have bounced off like a flat stone from water. Too steep an angle would have subjected the craft to catastrophically high temperatures and forces. The correct path had however been computed long before Apollo 8 neared the Earth on the return journey, and all went perfectly. A double-skip manoeuvre to slow the craft down during re-entry actually resulted in a 30,000ft gain in altitude before the command module descended into the atmosphere again. At T+147hr the three astronauts made contact with their native planet once more, splashing down in the Pacific Ocean at 8°N 165°W, four miles from the carrier USS *Yorktown*. It took 88min for the astronauts to reach the deck of the carrier after being picked up by helicopter; this proved to be the longest recovery time in the programme. The $310 million, 6,000,000-mile flight was over and America was looking to 1969 – the year of the Moon.

Soyuz 4   January 14, 1969   Flight 42
Soyuz 5   January 15, 1969   Flight 43

Name: Soyuz 4
Sequence: 42nd astro-flight, 29th spaceflight, 27th Earth orbit
Launch date: January 14, 1969
Launch site: Tyuratam, USSR
Launch vehicle: A2 (SL-4)
Flight type: Earth orbit and docking
Flight time: 2 days 23hr 20min 47sec
Spacecraft weight: 14,608lb
Crew: Lt-Col Vladimir Alexandrovich Shatalov, 42, Soviet Air Force, group flight commander
Name: Soyuz 5
Sequence: 43rd astro-flight, 30th spaceflight, 28th Earth orbit
Launch date: January 15, 1969
Launch site: Tyuratam, USSR
Launch vehicle: A2 (SL-4)
Flight type: Earth orbit and docking
Flight time: 3 days 0hr 54min 15sec
Spacecraft weight: 14,519lb
Crew: Lt-Col Boris Valentinovich Volynov, 34, Soviet Air Force, commander
*Alexei Stanislavovich Yeliseyev, 34, flight engineer
*Lt-Col Yevgeni Vasilyevich Khrunov, 35, Soviet Air Force, research engineer
*Landed in Soyuz 4

Left to right Alexei Yeliseyev, Soyuz 5 flight engineer; Yevgeni Khrunov, Soyuz 5 research engineer; Vladimir Shatalov, Soyuz 4 commander; and Boris Volynov, Soyuz 5 commander. Apart from being the seventh and eighth space-walkers respectively, Yeliseyev and Khrunov were the first men to return to Earth in a craft other than the one in which they were launched (Novosti)

Vladimir Shatalov was wearing a fur hat, boots and a woollen flight suit when he arrived at the launch pad to fly Soyuz 4 into space on January 14, 1969. And in another indication of confidence in the Soyuz design, his launch from Tyuratam at 10.39 am was recorded on television and transmitted to the West shortly afterwards.

Born in Petropavlovsk on December 8, 1927, Shatalov became a cadet pilot in Kachimo shortly after the war and in 1949, at the age of 22, was assigned as an instructor pilot. He graduated from the Air Force Academy in Moscow before becoming assistant commander of a flight detachment and a squadron leader. Joining the cosmonaut team in 1963, Shatalov first served as Soyuz 3 back-up pilot.

Another launch was expected, and it came at 10.14 am the

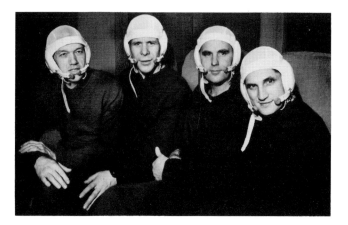

The clustered rocket motors of a Soyuz launch vehicle (Novosti)   *Launch of a Soyuz* (Novosti)

following day. On board Soyuz 5 were three crewmen. Commander was Boris Volynov, born on December 18, 1934, in Irkutsk. He went to school in Prokopyevsk and entered aviation college in 1956. Volynov was one of the youngest cosmonauts chosen in 1960 and served as back-up to Vostok 5 pilot Bykovsky. He graduated with a degree from the Air Force College in 1968.

Flight engineer was civilian Dr Alexei Yeliseyev. He was born on July 13, 1934, in Zhizdra and went to technical school in Moscow. A Master of Technical Sciences, he worked as a spacecraft designer before joining the cosmonaut team in 1966 as a member of the first group of civilians.

Lt-Col Yevgeni Khrunov was the research engineer on Soyuz 5. Born in Prudy on September 10, 1933, he attended agricultural college before joining the Soviet Air Force in 1953. He graduated with a science degree to become an "aviator-scientist" and joined the first cosmonaut team in 1960. As back-up pilot of Voskhod 2 he was a trained spacewalker.

Soyuz 4 had completed 33 orbits and Soyuz 5 a total of 17 when, on January 16, the automatic rendezvous sequence brought the craft to within 328ft of each other. A television camera was switched on and, 45min later, showed the two craft docking under Shatalov's manual control. As docking was achieved at 11.20 am Moscow time on January 16 the Soyuz 5 crew shouted: "You raped me!" They were subsequently reprimanded for their choice of words.

This success allowed the Soviet Union to claim that the first "experimental space station" (weighing 28,497lb) had been established, with four men in "one" craft. Soon Yeliseyev and Khrunov were sitting alongside Shatalov after making the first space transfer the hard way – by spacewalking. Wearing suits equipped with life-support systems, the cosmonauts had first entered the orbital module of Soyuz 5, closed the hatch leading to the flight cabin, depressurised the module, opened a hatch and scrambled along the "space train" with the aid of handrails. Within 37min Yeliseyev and Khrunov were inside the Soyuz 4 orbital module and an hour later hatches had been closed and the modules repressurised. The cosmonauts became the first space postmen, delivering mail to Shatalov. The two craft separated after 4hr 33min 49sec and the crews carried out independent work in the fields of geography, geology, navigation, medicine and radio communications.

Soyuz 4 touched down 25 miles north-west of Karaganda in a temperature of –35°C. The craft had made 48 orbits in 71hr 20min. Yeliseyev and Khrunov became the first men to return to Earth in a different spacecraft, clocking up personal times of 47hr 45min. Volnyov, in Soyuz 5, landed a day later 125 miles south-west of Kustanai after 49 orbits in 72hr 54min. The entire mission had lasted 96hr 30min and was probably prompted by the fact that the Americans were planning a similar exercise on Apollo 9.

Name: Apollo 9 (AS-504)
Sequence: 44th astro-flight, 31st spaceflight, 29th Earth orbit
Launch date: March 3, 1969
Launch site: Pad 39A, Kennedy Space Centre, Merritt Island, USA
Launch vehicle: Saturn V (SA-504)
Flight type: Earth orbit, rendezvous and docking
Flight time: 10 days 1hr 0 min 54sec
Spacecraft weight: 80,599lb (CM/SM/LM)
Crew: Col James Alton McDivitt, 39, USAF, commander
Col David Randolph Scott, 36, USAF, command module pilot
Russell Louis "Rusty" Schweickart, 33, lunar module pilot

The Apollo 9 crew take a breather during water egress training in preparation for their mission. Left to right: Rusty Schweickart, Dave Scott and James McDivitt (Nasa via Astro Information Service)

It fell to the three astronauts of Apollo 9 to demonstrate the integrity of the whole Apollo system and to show that the American dream of a man on the Moon by 1970 could still come true. One central question remained to be answered: could the lunar module (LM) fly in space, rendezvous and dock with the mother ship? In command of this mission was Gemini 4 veteran James McDivitt. His command module pilot was David Scott of Gemini 8, and the lunar module pilot – the man who would assist the commander to fly the LM – was Russell "Rusty" Schweickart, aged 33.

Brought up on a farm in Neptune, New Jersey, where he was born on October 25, 1935, Schweickart studied aeronautical engineering and received a science degree in 1956. He entered the USAF and flew for five years before studying astronautics and aeronautics at MIT, from which he graduated shortly before his selection with the Group 3 astronauts in October 1963. At MIT he studied upper-atmosphere physics, applied astronomy and star tracking. He was an interesting choice for what was to be a purely test-flying role.

The LM was 22ft 11in high and consisted of descent and ascent stages. The descent stage – 10ft 7in high, 14ft 1in wide and weighing 22,209lb – had four legs and footpads and a descent engine. On lunar flights it would be equipped with experiments. The 9,812lb ascent stage, which was to take off

from the Moon using the descent stage as a launch pad, was 12ft 4in high and 14ft 1in wide and housed the crew cabin and the ascent engine.

The Saturn V had been man-rated, the command and service modules had been round the Earth and the Moon with men aboard, but the lunar module, the machine on which a Moon landing ultimately depended, had never been test-flown by astronauts. It was indeed a make-or-break flight. The original launch day, February 28, passed quietly, with the three spacemen recovering from the colds that had stopped them boarding the spacecraft on time.

At 11 am on March 3, 1969, Saturn V SA-504, generating a total thrust of 7,700,000lb, rose steadily from Pad 39 and climbed into Earth orbit. Once in space, Scott separated the command module and service module (CM/SM) from the S-IVB. Encased in the stage was the craft on which so much depended. Scott turned the CM/SM around and nosed into the S-IVB's gaping mouth, docked with the lunar module and extracted it, demonstrating that the key transposition and docking manoeuvre was practicable. The S-IVB was then commanded to fire its engine, sending it on a lonely journey into an eternity of solar orbit. The astronauts watched it speed away: "It's just like a bright star disappearing in the distance," said Scott. Most of the next 24hr were spent checking the two docked spacecraft, with their combined weight of 81,257lb, and in eating and sleeping.

Schweickart was scheduled to test the Lunar Extra-vehicular Activity (LEVA) spacesuit and Portable Life Support System (PLSS) backpack during a two-hour spacewalk from the LM during which he was to have transferred from LM to CM, but this was postponed when he suffered spacesickness. By the fourth day he had recovered enough to stand outside the LM, at the top of the stairs down which he would have climbed to the surface if he had been on the Moon. He stayed put by fitting his feet securely into the "golden slipper" restraints on the LM porch. While Schweickart tested his suit and took still and motion pictures, Scott opened the hatch of the CM and leaned out to retrieve materials samples and experiments on the side of the spacecraft. Each EVA lasted about 46min and Schweickart seemed the better for his walk outside.

The fifth day began with McDivitt and Schweickart entering the LM, nicknamed *Spider*, powering up its systems and undocking from the CM, nicknamed *Gumdrop*. After flying around the CM to allow Scott to look the LM over, McDivitt fired the descent-stage engine to simulate the first stages of a lunar landing. This burn put the LM into an orbit 13 miles above the CM, and after two hours the two modules were 50 miles apart. They drifted closer to each other once more and McDivitt fired the descent engine again to simulate the beginning of the flight to the lunar surface. This was followed by the separation of the spindly-legged descent stage and the first use of the ascent-stage engine, which was fired to simulate a lunar take-off. The radar and computer on board the LM were then set to work calculating the various burns needed to achieve rendezous with Scott in *Gumdrop* 100 miles away. Everything worked perfectly and as Scott spotted the curiously shaped ascent stage he shouted: "You're the biggest, friendliest, funniest looking spider I've ever seen." The docking, performed by McDivitt, was difficult: "That wasn't a docking, that was an eye test," he said. It was later decided that on subsequent missions the CMP would perform the manoeuvre.

The mission proceeded quietly after this, with the crew performing valuable experiments; a highly efficient Earth-resources camera was of particular interest. As the end of the flight approached an extra orbit was flown, for the first time on an American mission, to ensure that Apollo 9 would miss heavy seas 195 miles south-west of Bermuda. The SPS engine was fired for 11.8sec to reduce speed by 221mph and induce re-entry, and the CM splashed down in the Atlantic three miles from USS *Guadalcanal* 535 miles south of Bermuda. Apollo 9 had covered 151 revolutions and 4,350,000 miles in 10 days 1hr 53sec, and America was "go" for the Moon.

The first manned lunar module in space: Spider *viewed from* Gumdrop *before the first firing of the descent engine. The probes protruding from the footpads of the LM were used during the lunar landings to signal contact with the surface and command engine shutdown* (Nasa via Astro Information Service)

Apollo 10   May 18, 1969   Flight 45

Name: Apollo 10 (AS-505)
Sequence: 45th astro-flight, 32nd spaceflight, 30th Earth orbit, 2nd lunar flight, 2nd lunar orbit
Launch date: May 18, 1969
Launch site: Pad 39B, Kennedy Space Centre, Merritt Island, USA
Launch vehicle: Saturn V (SA-505)
Flight type: Lunar orbit, rendezvous and docking
Flight time: 8 days 0hr 3min 23sec
Spacecraft weight: 94,512lb (CM/SM/LM)
Crew: Col Thomas Patten Stafford Jr, 38, USAF, commander
Cdr John Watts Young Jr, 38, USN, command module pilot
Cdr Eugene Andrew Cernan, 35, USN, lunar module pilot

Smiles all round aboard the recovery ship USS Princeton *as the Apollo 10 crew celebrate their return to Earth. Left to right: Eugene Cernan, John Young and Tom Stafford* (Nasa via Astro Information Service)

When the Apollo 10 crew was named in October 1968, during the flight of Apollo 7, they were assigned a vehicle incorporating an LM fully capable of landing on the Moon. If all had gone to plan, two of them would have been the first men on the Moon. But then production difficulties led to Apollo 10 being given an LM capable of everything except a lunar landing. Accordingly, their mission was to be a complete dress rehearsal, all bar a touchdown on the Moon's surface. The astronauts named their craft after the famous cartoon characters Charlie Brown and Snoopy, a choice that characterised the three men. Tom Stafford and John Young went into Apollo service early and were tentatively assigned flights. Eugene Cernan joined them later after his stint as Gemini 12 back-up pilot, and eventually the three became the Apollo 7 back-up crew.

What was "definitely the riskiest of any mission put together," as Stafford described it, began from Pad 39B at

Sixty miles above the Moon, command module pilot John Young is the first man to fly solo in lunar orbit (Nasa via Astro Information Service)

12.49 pm on May 18, with Apollo 10 riding atop the fifth Saturn V and its 7,680,000lb of thrust. The S-IVB lit up over the Pacific Ocean on the second Earth orbit and Stafford shouted "We're on the way!" The crew described this phase of the ascent as a "rocky ride". Television on Earth showed colour views of the planet and the transposition and docking of command module *Charlie Brown* and lunar module *Snoopy*. The 19 television spectaculars during the flight were fascinating, friendly and not only extended but often unscheduled.

Apollo 10 entered lunar orbit on May 21, and after television shows, photographic operations and a final pre-

flight examination of *Snoopy* it was time for a low swoop over the Moon. *Snoopy* was to go half-way towards a lunar landing while Stafford and Cernan closely monitored its performance and observed the area where the Apollo 11 crew were to land. The craft separated on the 12th orbit and *Snoopy*'s descent engine was fired for 27.4sec. The LM swooped in low over the Sea of Tranquillity at a height of nine miles. "Houston, Houston, this is *Snoopy*. We is low. We is down among 'em Charlie!," Cernan shouted to capsule communicator Charles Duke. When Duke asked Cernan for a report about the engine firing Cernan replied: "Yeah, just let me get my breath!" A few seconds later the two astronauts saw planet Earth rising serenely over the horizon. Cernan could not contain himself: "Aw Charlie, we just saw Earthrise and it's got to be magnificent!" It was, as photos later showed.

As the duo swept over the edge of the Sea of Tranquillity, Stafford saw so many boulders that he remarked: "There's enough boulders here to fill up Galveston Bay!" When he saw the Apollo 11 target area he said it looked like "very smooth, wet clay, like a dry riverbed in New Mexico or Arizona."

Then, as the descent stage was jettisoned and the ascent stage readied for firing, euphoria nearly gave way to disaster. The ascent stage went into a spin and it took Stafford 8sec to bring it under control. Cernan obviously thought that the craft was going to crash: "Son of a bitch! There's something wrong with the gyro . . . I don't know what the hell that was babe . . . I'll tell you, there was a moment there, Tom . . ." The shaken astronaut didn't continue, but everyone knew what he meant. Turning his attention to the pending engine firing, he said: "Let's worry about it when we make this burn . . . OK Charlie, I think we got all our marbles . . . We're sure coming

down to that ground, I'll tell you . . ." Then, referring once more to the gyrations: "I don't know what happened there but I hope we never find it again."

A successful ascent-engine burn followed and an obviously relieved Cernan said: "Baby, that made me feel better." Then came the slow ascent to *Charlie Brown* and a rendezvous with the quiet John Young, who had meanwhile become the first man to fly solo around the Moon. "Hello Houston, *Snoopy* and *Charlie Brown* are hugging each other," said Stafford, announcing a safe docking. "Man! We is back home. That rendezvous was the best thing we ever had," said Cernan.

Charlie Brown lit up its engine and leaped out of the Moon's grasp after 31 orbits in 61hr 34min. Apollo 10 came home 7,000 yards from USS *Princeton* at 165° W 5° S after a flight of 830,000 miles in eight days. The Apollo bill had gone up another $350 million and Stafford, Young and Cernan had done everything – except make history.

Apollo 11   July 16, 1969   Flight 46

Name: Apollo 11 (AS-506)
Sequence: 46th astro-flight, 33rd spaceflight, 31st Earth orbit, 3rd lunar flight, 3rd lunar orbit, 1st lunar landing
Launch date: July 16, 1969
Launch site: Pad 39A, Kennedy Space Centre, Merritt Island, USA
Launch vehicle: Saturn V (SA-506)
Flight type: Lunar landing
Flight time: 8 days 3hr 18min 35sec
Spacecraft weight: 96,715lb (CM/SM/LM)
Crew: Neil Alden Armstrong, 38, commander
Lt-Col Michael Collins, 38, USAF, command module pilot
Col Edwin Eugene "Buzz" Aldrin, 39, USAF, lunar module pilot

On January 9, 1969, Nasa made the long awaited announcement of the names of the crew who would attempt the first landing on the Moon. After the delays in the development of the lunar module it had become obvious that the first landing attempt would be made by Apollo 11 rather than its predecessor. The Apollo 11 crew weren't in any way special among the astronaut corps: it's just that Neil Armstrong, Michael Collins and Edwin Aldrin were next in the queue.

Armstrong, who had served as Gemini 11 back-up commander, came into the Moon programme as back-up commander of Apollo 9, which became Apollo 8 in August 1968. It was at this point that it was first predicted that Armstrong would be the first man on the Moon, the general practice in Gemini and Apollo being to pick as prime crew the next available unassigned back-up crew. The other members of the Apollo 8 back-up crew were Aldrin and a new astronaut, Fred Haise. But by the time the Apollo 11 crew was selected Michael Collins had fully recovered from the neck and back injury which had caused him to drop out of Apollo 8 and to be replaced by James Lovell.

The objective of the $355 million Apollo 11 mission was

Above *The Apollo 11 crew in July 1969* ***Below*** *Mission accomplished: one of Man's first footprints on the Moon* (Nasa)

quite simply "to land two men on the Moon and return them safely to Earth," so achieving the national goal set by President Kennedy. The flight began at 9.32 am as the now tried and trusted Saturn V rose ponderously from Pad 39A to the cheers, waves, wishes and prayers of a million spectators, including 3,000 pressmen, and 600 million television viewers around the world. At T+11min 42sec the S-IVB shut down, putting Apollo 11 into a 114/116-mile parking orbit. After 1¾ orbits the 203,779lb-thrust S-IVB engine restarted and fired for 5min 47sec, taking the spacecraft to a speed of 24,545mph. The crew were very quiet; during the burn Armstrong had said only: "Ignition ... We have no complaints with any of the three stages on that ride. It was beautiful ... Shutdown."

Apollo 11 became the command module *Columbia* and lunar module *Eagle* after the transposition and docking manoeuvre to extract the lander from the S-IVB. A quiet, subdued journey, which included some live television, then followed. Behind the Moon the astronauts fired the SPS engine for 357sec and safely entered a preliminary orbit. As *Columbia* came around the other side of the Moon and into contact with Earth once more Armstrong described the firing: "It was like perfect. Everything looks good up here." The first Apollo 11 television transmission from the Moon was then shown, with Armstrong commenting on the appearance of one of the Moon's seas: "The Sea of Fertility doesn't look very fertile to me!" The view from orbit was well worth the trip, he reported.

Eagle and *Columbia* separated on the far side of the Moon at T+100hr 12min, and as the LM came into Earth view Armstrong reported: "The *Eagle* has wings." On the next revolution the descent engine of Eagle fired for 30sec on the far side to put the LM into its descent orbit. As *Columbia* emerged from the Moon's shadow Collins gave his view of events so far: "Listen babe, everything's just going swimmingly – beautiful."

As *Eagle* came into sight Armstrong reported that his orbit was 57.2 miles by 9.1 miles. Aldrin could be heard reading off the checklist for the powered descent initiation burn. As the craft reached 9.1 miles above the lunar surface, the lowest point of its orbit, the engine was ignited to fire for 756sec. Communications deteriorated at this point, and it was some time before Duke could be heard saying: "OK *Eagle*, we got you now, everything's go."

The lunar module was at 47,000ft, 2min 11sec into the burn, when an alarm light flashed in the cockpit to indicate that the computer was being overloaded with commands. But Mission Control told the astronauts that there was no need for alarm: "You are go to continue powered descent," Duke called twice. Aldrin reported that he could see the Earth out of his window as the craft turned around for the final approach to the Sea of Tranquillity. Another alarm came on and Armstrong asked Duke if he was still "go" to descend further: "Give me a reading on the programme alarm." Doubts about *Eagle's* fitness to continue began to grow. But then Steve Bales, the young engineer in charge of the suite of computer monitoring equipment, urged that the astronauts be instructed to ignore the alarm and carry on. A brave decision was taken, and Duke told *Eagle*: "You are go, you are go". At this point the LM was at 21,000ft and descending at 12,000ft/min. As the LM reached 4,200ft Duke gave another go-ahead: "You are go for landing". Then Aldrin announced another computer alarm: "12.01 alarm." Duke responded: "Hang tight, you're go." The descent continued through 1,400ft.

Armstrong looked out of his window and saw, directly ahead, a huge crater strewn with rocks. With *Eagle* looking set to touch down in this hazardous jumble, the Apollo 11 commander took control, stretching the flightpath in search of a smoother landing area.

Aldrin called out instrument indications of flightpath angle, altitude, and vertical and horizontal speed in feet per second: "35° ... 750 ... coming down in 23 ... 700 down 33° ... 400 ... down at 9 ... 8 ... forward ... 70 ... 3½ down ... 15 forward ... 200ft ... 4½ down ... 5½ ... 9 forward ... 100ft ... 3½ down ... 9 forward."

"60 seconds," Duke shouted, warning that there was just a minute of descent-engine fuel remaining. Aldrin's talkdown continued: "Down 2½ ... forward ... forward ... 2½ ... picking up some dust .. faint shadows .. drifting to the right a little." Duke interjected: "30 seconds." Then came the first of a series of historic transmissions from *Eagle*: "Drifting right ... contact light!" Armstrong waited a second and switched off the engine as the craft settled. Aldrin said: "OK, engine stop." It was 9.18 pm BST on July 20, 1969, and Man was on the Moon.

Aldrin squeezes through the exit hatch before becoming the second man on the Moon (Nasa)

The far side of the Moon pictured from Apollo 11 in lunar orbit (Nasa)

"We copy you down, *Eagle*," said Duke. Armstrong came on the line and quietly uttered these historic words: "Houston. Tranquillity Base here. The *Eagle* has landed." A breathless Duke replied: "Roger Tranquillity, we copy you on the ground. You've got a bunch of guys about to turn blue. We're breathing again. Thanks a lot." *Eagle* had landed four miles downrange from the predicted touchdown point at 0° 41′15″N 23° 26′E.

Four hours after landing the astronauts had donned the Portable Life Support System backpacks and bulky suits which would keep them alive in the harshest environment ever braved by men: hard vacuum, temperatures varying from –279°F in the shade to 240°F in the glare of the Sun, and

unshielded solar radiation. Armstrong squeezed himself feet-first out of a small hatchway and stepped on to the "porch" and then on to the first of nine rungs on the ladder to history. On the way down he pulled a lanyard which released the package of instruments that the astronauts would be using on the surface and also exposed a black-and-white television camera. At first the ghostly picture was upside down, showing Armstrong's legs climbing down the remaining rungs. Then he jumped 3ft onto the landing pad and his whole body was visible. He began his report by making a few observations about the immediate area: "The LM footpads are only depressed in the surface about one or two inches, although the surface appears to be very, very fine-grained as you get close to it; it's almost like powder. Now and then its very fine." A brief silence followed and then he said: "I'm going to step off the LM now." The historic moment had arrived.

Armstrong placed his left foot on the surface of the Sea of Tranquillity at 3.56 am BST on July 21, 1969. His no doubt carefully considered words have since been quoted many times: "That's one small step for man. One giant leap for mankind." (The official version had the word "a" added before "man", but in fact Armstrong did not say that.) Posterity satisfied, the first man on the Moon got down to a businesslike description of his surroundings: "The surface is fine and powdery. I can pick it up loosely with my toe. It does adhere in fine layers like powdered charcoal to the sole and sides of my boots. I only go in a small fraction of an inch. Maybe an eighth of an inch – but I can see the footprints of my boots and the treads in fine sandy particles."

Turning his attention to the sensations of movement in lunar gravity, he observed: "There seems to be no difficulty in moving around. As we suspected, it's even perhaps easier than the simulations at 1/6th g that we performed in periods of simulations on the ground. It's actually no trouble to walk around. OK, the descent engine did not leave a crater of any size .. It's about one foot clear from the surface. We're essentially on a level plain here."

It was now time to acquire a contingency sample of the surface, so that if the astronauts had to depart in a hurry the scientists would at least have a small quantity to examine. Using a scoop, Armstrong picked up some dust and put it into a special pocket in his spacesuit leg.

Then came Aldrin's turn to set foot on the Moon. What he saw prompted him to the description "magnificent desolation," echoing Armstrong's earlier assertion that the surface had a stark beauty all of its own. The two men then unveiled a plaque fixed to one of the landing legs. Armstrong read out the message it bore: "Here men from the planet Earth first set foot upon the Moon July 1969 AD. We came in peace for all mankind." The astronauts erected a flag and laid out more ceremonial objects, including a microdot message from world leaders and a memorial to the dead astronauts and cosmonauts. They also carried out an embarrassing public telephone conversation with President Nixon. By now the scientists were losing patience with the stage-managed ceremonies and the lack of emphasis on exploration.

But in truth the experiments and rock-gathering were something of a bonus on this flight, the object of which was simply to prove that man could land on the Moon. In 2hr 31min 40sec (hatch open to hatch closed) of intense activity the astronauts set up two instruments, a laser reflector and a seismometer, collected 48.5lb of rocks and evaluated methods of walking on the Moon. Aldrin had little difficulty in walking, running and making quick stops and turns in the lunar dust, which at first he had described as slippery. He also found time to try some comical-looking "kangaroo hops." Armstrong at one point moved as far as 300ft from the lunar lander.

After 21hr 36min on the Moon, during which Armstrong had walked outside for 2hr 14min and Aldrin for 1hr 33min, the two men set off home by firing the ascent engine for 435sec, entering orbit, manoeuvring to a rendezvous with Michael Collins aboard *Columbia*, and docking with the command module. After transferring to *Columbia* with their lunar spoils they jettisoned the ascent stage and prepared for the crucial SPS burn that would put them on course for Earth. Passing behind the Moon, they fired the SPS for 2min 29sec and *Columbia* was Earthwards-bound after circling the Moon 30 times in 59hr 30min. As *Columbia* came into view at the end of the SPS burn Armstrong said to capcom Duke: "Open the LRL doors, Charlie!" The LRL was the Lunar Receiving Laboratory, where the astronauts and their rocks would remain for three weeks after landing in order to avoid any possibility of contaminating the Earth with unsuspected lunar organisms.

At T+195hr 18min Apollo 11 splashed into the Pacific 210 miles from Johnson Island, at 169° W 13° N, and 13 miles from USS *Hornet*, on which President Nixon waited to greet them. Wearing special biological isolation garments, they were picked up by helicopter and flown to the carrier and a waiting "quarantine container". The first men to set foot on another heavenly body would have to wait a little longer before receiving the congratulations of the world.

Soyuz 6   October 11, 1969   Flight 47
Soyuz 7   October 12, 1969   Flight 48
Soyuz 8   October 13, 1969   Flight 49

Name: Soyuz 6
Sequence: 47th astro-flight, 34th spaceflight, 32nd Earth orbit
Launch date: October 11, 1969
Launch site: Tyuratam, USSR
Launch vehicle: A2 (SL-4)
Flight type: Earth orbit
Flight time: 4 days 22hr 42min 47sec
Spacecraft weight: 14,502lb
Crew: Lt-Col Georgi Stepanovich Shonin, 34, Soviet Air Force, commander
Valeri Nikolayevich Kubasov, 34, flight engineer

Name: Soyuz 7
Sequence: 48th astro-flight, 35th spaceflight, 33rd Earth orbit
Launch date: October 12, 1969
Launch site: Tyuratam, USSR
Launch vehicle: A2 (SL-4)
Flight type: Earth orbit
Flight time: 4 days 22hr 40min 23sec
Spacecraft weight: 14,486lb
Crew: Lt-Col Anatoli Vasilyevich Filipchenko, 41, Soviet Air Force, commander
Vladislav Nikolayevich Volkov, 33, flight engineer
Lt-Col Viktor Vasilyevich Gorbatko, 34, Soviet Air Force, research engineer

Name: Soyuz 8
Sequence: 49th astro-flight, 36th spaceflight, 34th Earth orbit
Launch date: October 13, 1969
Launch site: Tyuratam, USSR
Launch vehicle: A2 (SL-4)
Flight type: Earth orbit
Flight time: 4 days 22hr 50min 49sec
Spacecraft weight: 14,654lb
Crew: Col Vladimir Alexandrovich Shatalov, 42, Soviet Air Force, group flight commander
Alexei Stanislavovich Yeliseyev, 35, flight engineer

Good view of the Soyuz complex at Tyuratam as Soyuz 8 is prepared for launch. Soviet safety measures are not as stringent as Nasa's: technicians can be seen walking near the fully fuelled launcher (Novosti)

When Soyuz 6 soared into orbit with two cosmonauts on board at 2.10 pm on October 11, 1969, it was thought by experts in the West to be the beginning of a very special flight. To a certain extent they proved to be right. By the time Soyuz 6 had been in orbit for two days it was flying around the Earth with two sister ships, and seven men were in space. Soyuz 7, manned by three new cosmonauts, had followed Soyuz 6 a day later, and Soyuz 8, with veterans Shatalov and Yeliseyev aboard, joined them after another 24hr. The five new spacemen were Shonin and Kubasov of Soyuz 6 and Filipchenko, Volkov and Gorbatko of Soyuz 7.

Georgi Shonin, Soyuz 6 commander, was born in Rovenki in the Ukraine on August 3, 1935. He became an aviation cadet at the age of 15 and went to the Soviet Naval College. In 1957 Shonin received pilot's wings and was assigned to various units in the Baltic and Northern fleets. He joined the first group of cosmonauts in 1960.

Soyuz 6 flight engineer was Valeri Kubasov, born on January 7, 1935, in Vyasmki. He went to the Moscow Aviation

*Quick march to the medals at Moscow Airport after the strange "troika" mission of Soyuz 6, 7 and 8. **Left to right** Shatalov, Shonin, Filipchenko, Kubasov, Gorbatko, Yeliseyev and Volkov (Novosti)*

Left to right *Filipchenko, Volkov and Gorbatko after the flight*

The three descent modules are inspected by the crews (Novosti)

Institute and graduated in 1958 as an aircraft engineer. Later he graduated with a science degree. Kubasov joined the cosmonauts in 1966 and was the back-up flight engineer for Soyuz 5.

The commander of Soyuz 7, Anatoli Filipchenko, was born in Davydovka on February 26, 1928. He worked as a lathe operator before attending military school and joining the Soviet Air Force. Filipchenko joined the cosmonauts in 1963 and served as back-up Soyuz 4 commander.

Soyuz 7 flight engineer was civilian Vladislav Volkov, born on November 23, 1935, in Moscow. He became an aircraft designer and went to the Aviation Institute in Moscow before joining the cosmonauts in 1966.

Viktor Gorbatko was the Soyuz 7 research engineer. He joined the cosmonaut team in 1960 and served as back-up research engineer for Soyuz 5. He was born on December 3, 1934, in Kuban and joined the Air Force at the age of 19. He was lucky to be making a flight, an irregular heartbeat and a parachute accident having taken him off the active roster for some time.

Soyuz 6 was equipped as a miniature space station, with its orbital module packed full of experiment packages, including a unique welding set called Vulkan. This equipment had been fitted at the expense of a docking mechanism, ruling out a direct link-up with the other craft. Soyuz 7 and 8, on the other hand, did have docking equipment. But the nearest they got to each other was 1,600ft, observed from a distance of about a mile by Soyuz 6. It is thought that a manual control problem prevented a docking; if this had been achieved, a number of the cosmonauts would have changed spacecraft by crawling through the docking tunnel.

The three craft kept station for about a day before each went its own way to carry out a wide variety of manoeuvres and navigation, orientation, science and astronomical experiments, the most interesting of which was Vulkan. On

Soyuz 6's 77th orbit the orbital module was depressurised and the welding set ran through three processes – electron beam, fusible electrode and compressed arc – in the resulting combination of vacuum and zero g. The experiment was controlled by Kubasov from the flight cabin, and samples were returned to Earth.

The joint flight included a total of 31 change-of-orbit manoeuvres, the last of which initiated the crafts' re-entries. Soyuz 6 was first down, landing 112 miles north-west of Karaganda on October 17 after 80 orbits in four days 22hr 42min. Soyuz 7 came down a day later 96 miles north-west of Karaganda, and Soyuz 8 arrived 90 miles north of Karaganda on October 19 after 80 orbits and just one minute less than Soyuz 6.

Well, what was it all about? It is possible that if the Soviet space programme had had a few more lucky breaks the Russians could have sent a man round the Moon or even landed him on it before Apollo 11. The necessary flight modules were all based on the Soyuz design and had been tested under the guise of Polyot, Cosmos and Zond. To get to the Moon for a landing would presumably have meant a docking in Earth orbit of two Soyuz, the transfer of cosmonauts and then their despatch out of Earth orbit.

Perhaps failures of spacecraft, Soyuz included, and launch vehicles, plus the success of the Apollo programme, caused the abandonment of the Russian manned lunar programme, leaving the hardware available before it had anywhere to go in the form of the planned space station. Hence the strange, unique flights of Soyuz 3, 4, 5, 6, 7 and 8.

Another analysis attibutes this sequence of flights to the initial failure to launch a Salyut space station, leaving the planned Soyuz ferry craft unemployed. Soyuz 6 was probably to have been a solo engineering flight, delayed due to pressure to beat Apollo 11 by launching Luna 15 on a lunar sample-return mission. Soyuz 7 was possibly due to fly a test rendezvous with a boilerplate Salyut which subsequently failed. This led to a plan for Soyuz 7 to dock with a later Salyut, with Soyuz 6 in attendance. Then that Salyut also ran into problems. Finally the Russians wanted to fly Soyuz 8 before the winter set in, so in desperation they orbited all three together, seeking a few crumbs of propaganda comfort from what remains to this day the largest group flight ever attempted.

The Salyut jinx was to afflict one more Soyuz mission, the failure of yet another space station reducing Soyuz 9 of June 1970 to a solo science mission.

Apollo 12   November 14, 1969   Flight 50

Name: Apollo 12 (AS-507)
Sequence: 50th astro-flight, 37th spaceflight, 35th Earth orbit, 4th lunar flight, 4th lunar orbit, 2nd lunar landing
Launch date: November 14, 1969
Launch site: Pad 39A, Kennedy Space Centre, Merritt Island, USA
Launch vehicle: Saturn V (SA-507)
Flight type: Lunar landing
Flight time: 10 days 4hr 36min 25sec
Spacecraft weight: 96,812lb (CM/SM/LM)
Crew: Cdr Charles "Pete" Conrad Jr, 39, USN, commander
Cdr Richard Francis Gordon Jr, 40, USN, command module pilot
Cdr Alan Lavern Bean, 37, USN, lunar module pilot

The task of Apollo 12 was not just to land on the Moon but to make a pinpoint arrival 1,000 yards from Surveyor 3, the unmanned lunar lander which had touched down in the Ocean of Storms in April 1967. Success would clear the way for further precision landings in other areas of scientific interest.

The Apollo 12 crew were all Navy commanders, making their third, second and first flights into space respectively. They were Charles Conrad, Richard Gordon and new man Alan Bean. Born in Wheeler, Texas, on March 15, 1932, Bean graduated in aeronautical engineering from the University of Texas in 1955. He joined the Navy, graduated from that service's test pilot school and joined the third group of Nasa astronauts in October 1963. Bean was the Gemini 10 back-up command pilot.

On November 14, 1969, conditions at the Kennedy Space Centre were dark, wet and thundery, with the launch pad hardly visible from the press site. Indeed, so marginal was the weather that the decision to go ahead with the launch at

The Apollo 12 crew pictured while training as the Apollo 9 back-up team. Left to right: Charles Conrad, Dick Gordon and Alan Bean (Nasa via Astro Information Service)

11.22 am could well have been influenced by the presence of President Nixon at the Cape and by the need for political support for Apollo now that the Moon landing had been achieved. The size of the risk was soon revealed when, at T+36sec, bolts of lightning hit the Saturn. From Apollo 12, after the strike, came the report: "OK we just lost platform, gang. I don't know what happened here – we just had everything in the world drop out." Hit by a massive power surge as a result of the lightning strikes, the spacecraft's circuit-breakers had tripped, temporarily shutting down vital systems. Order was soon restored, however, and as the S-II second stage ignited Conrad commented on the incident: "Staging. We got a good S-II. We had our problems here . . . I'm not sure that we didn't get hit by lightning." The suggestion was denied by Nasa officials at first but was later given credence when, towards the end of the mission, Conrad saw scorch marks on the side of the command

Apollo 12's Saturn V ascends in a thunderstorm. The spacecraft was struck by lightning half a minute later. Apollo 12 was the 50th flight above 50 miles (Nasa)

module. "I think we need to do a little more all-weather testing," chipped in Richard Gordon during the ascent.

Insertion into Earth orbit was achieved without further incident and then *Yankee Clipper*, the command module, and lunar module *Intrepid* were sent on their way to the Moon by a 5min 45sec S-IVB burn. A later mid-course correction put Apollo 12 into a "hybrid" trajectory, one that did not guarantee a return to Earth. Previous flights had followed a free-return trajectory, which meant that if nothing else happened the spacecraft would loop the Moon and come back to Earth. The Apollo 12 path was necessitated by the location of the landing site.

The now-traditional en route television transmissions were chatty and friendly, and Apollo 12 went into initial lunar orbit on November 18. After more television and a checkout of *Intrepid*, Conrad and Bean said goodbye to Gordon and sped down to the Moon on November 19. The commander had a set of craters to aim for during the latter part of the landing, and he could hardly believe it when he saw how well the guidance system had worked: "I think I see my crater. I'm not sure . . . There it is! There it is! Oh my gosh! Right down the middle of the road! I can't believe it! Amazing! Fantastic!"

Bean read out the key instrument indications and enthusiastically encouraged Conrad down: "1,200ft . . . 600ft . . . 400ft . . . You got loads of gas . . . Come on down Pete . . . 180ft . . . 96ft, slow down descent rate . . . 60ft . . . 50ft . . . Watch for the dust . . . 42ft . . . 40ft . . . Lots of dust . . . 30ft . . . Plenty of gas . . . He's got it made . . . Come on in there . . . 24ft . . . Contact light!" The time was 7.54 am GMT and *Intrepid* had landed at 3° 11′S 23° 23′W just 553ft from Surveyor 3, with only 23sec of fuel remaining.

After suiting up the diminutive Conrad emerged from *Intrepid* and jumped the last three feet from the ladder to the surface: "Whoopee! Man, that may have been a small one for Neil but that's a long one for me," he joked. When Alan Bean joined him on the surface Conrad greeted his Navy colleague with "Welcome aboard."

For the first time colour television pictures from the surface were beamed to a worldwide audience. But Bean then spoiled the fun when, after taking the camera out of *Intrepid*'s side to fix it on a tripod away from the lunar module, he panned through 360°. The Sun proved to be rather too near the horizon for the camera's comfort, however, and the overloaded lens stopped working. "We did not know the Sun would hurt the TV," he said later. The astronauts had a hefty schedule ahead of them and could not waste time trying to fix the camera, although Bean did try to rectify matters in a very human way by hitting it with a hammer.

Conrad erected the second American flag to "fly" on the surface of the Moon, and the two humming and whistling astronauts were ready to tackle the first experiments. Their main job on the 3hr 56min first walk was to deploy the nuclear generator-powered Apollo Lunar Surface Experiment Package (ALSEP), which immediately began to send back information to Earth. The spacemen collected rock samples, giving a running commentary designed to help Mission Control to follow their movements and, perhaps even more important, to engage the interest of the television audience. Public enthusiasm for Apollo was now on the wane, and the astronauts were as aware anybody of the effect this might have on the future of the programme.

A walk lasting 3hr 49min the following day featured the astronauts' meeting with "good old Surveyor," as Conrad called the spacecraft. After a long, two-mile, circular tour of the area they came to the crater in which Surveyor 3 had

landed. Bean removed parts of the spacecraft to permit analysis of the effects of many months of exposure to the lunar environment.

Some 31hr 31min after landing the ascent stage lifted off, carrying 74.7lb of rock, pieces of Surveyor 3 and two spacemen with 7hr 24min (Conrad) and 6hr 17min (Bean) of moonwalking under their belts. The latter phases of the rendezvous with Gordon in *Yankee Clipper* made spectacular viewing for the television audience on Earth. The pictures were so good that viewers could see the faces of *Intrepid*'s crew as the craft came in for docking.

Conrad said later that he and Bean had performed the first space streak. Gordon had refused to allow their dusty spacesuits into his command module, so the two moonwalkers had floated naked from the ascent stage into the command module. Later, when *Intrepid* was jettisoned, they were wearing only their headsets. The abandoned ascent stage, deliberately directed towards the lunar surface, struck with such force that it caused a 30min moonquake which registered on the seismometer deployed by the astronauts.

Further work on Gordon's intensive lunar orbital science and photographic programme occupied the next few orbits, and then the crew fired the SPS engine and headed home. *Yankee Clipper* had spent 45 revolutions and 88hr 56min in lunar orbit. Splashdown came at T+244hr 36min 25sec in the Pacific near USS *Hornet* at 15°S 165°W.

When the astronauts looked at the photographs they had taken during the moonwalks they realised that they couldn't remember who was who in many of the pictures. As a result, it was decided that on future Apollo flights the commander would carry indentifying red strips on his arms and helmet.

When subsequently examining the photos they took on the Moon, Conrad and Bean had difficulty remembering who was whom, as in this picture of one of them deploying the ALSEP. Nasa subsequently decided that in future Apollo commanders would wear distinctive red stripes on their spacesuits (Nasa via Astro Information Service)

Apollo 13   April 11, 1970   Flight 51

Name: Apollo 13 (AS-508)
Sequence: 51st astro-flight, 38th spaceflight, 36th Earth orbit, 5th lunar flight
Launch date: April 11, 1970
Launch site: Pad 39A, Kennedy Space Centre, Merritt Island, USA
Launch vehicle: Saturn V (SA-508)
Flight type: Lunar fly-by
Flight time: 5 days 22hr 54min 41sec
Spacecraft weight: 96,964lb (CM/SM/LM)
Crew: Capt James Arthur Lovell Jr, 42, USN, commander
John Leonard "Jack" Swigert Jr, 39, command module pilot
Fred Wallace Haise Jr, 36, lunar module pilot

After separation of the command module from the service module the Apollo 13 crew could see for the first time the extent of the damage caused by the explosion, which had torn out an entire side panel (Nasa)

Public indifference to the Moon landings had set in by the time Apollo 13 rose majestically from Pad 39A at 2.13 pm on April 11, 1970. Nobody, it seemed, was interested in space anymore. Three days later, just as the mission seemed to be running on rails, the world received a sudden and violent reminder of the perils of spaceflight.

Apollo 13 was scheduled to make a tricky landing in the foothills of Fra Mauro, a few hundred miles from the Apollo 12 site. The crew consisted of one of the world's most experienced spacemen and two rookies. James Lovell, having acted as Apollo 11 back-up commander, was embarking on his fourth journey into space and his second to the Moon, and preparing for his first lunar walk. The lunar module pilot was Fred Haise, born in Biloxi, Mississippi, on November 14, 1933. By the time of his selection as an astronaut in April 1966 Haise had gained a science degree and flown jet aircraft for the US Navy, USAF, US Marine Corps, Nasa and the Oklahoma Air National Guard. His selection as Apollo 13 lunar module pilot followed service in this position on the back-up crews for Apollo 8 and 11.

The command module pilot was to have been Thomas "Ken" Mattingly, aged 34, who joined the Nasa astronaut pool in April 1966. But then the Apollo 13 jinx made its first appearance. A few days before the launch – which had already been put back from March 12 – one of the back-ups, Charles Duke, visited a house in which one of the children was developing the first stages of German measles. Duke continued training with the prime and back-up crews, not realising that he had caught the complaint. When he ended up in bed, doctors quickly studied the other Apollo 13 astronauts to see if they had an immunity to German measles. Mattingly didn't and he was dropped from the flight. His back-up, Jack Swigert, had been training for months in preparation for just such an eventuality, and so with two days to go he found himself working round the clock with Lovell and Haise to fit him into the Apollo 13 crew.

A civilian test pilot, Jack Swigert was born in Denver, Colorado, on August 30, 1931. His academic attainments included degrees in mechanical engineering in 1953, aerospace science in 1957 and business administration in 1967. Swigert flew in the USAF and was the only astronaut in his class to have flown in Korea. He became a civilian test pilot with North American and Pratt & Whitney before being selected as an astronaut in 1966. His appointment as Apollo 13 back-up command module pilot was preceded by service on the support crews for the Apollo 7 and 11 missions.

Apart from the premature shutdown of one of the five S-II engines during launch, obliging the remaining four to fire for an additional 34sec, the initial phase of the mission went like clockwork as command module *Odyssey* and lunar module *Aquarius* sped towards the Moon. The three crewmen sent back television transmissions which no US station regarded as newsworthy enough to broadcast live. Soon after one of these shows, when the craft was halfway to the Moon, trouble hit Apollo 13 and the news editors' dreams came true.

Inside Bay 4 of the service module was oxygen tank No 2, serial No 10024X-TA0009. Part of the fuel-cell system that provided electrical power to the spacecraft, this particular tank had originally been installed on Apollo 10 but had been removed after giving trouble. The tank was subsequently installed in Apollo 13 and during a pre-launch test on March 16, 1970, excessive electrical currents welded heater switches shut when liquid oxygen was boiled off in an effort to empty the tank, which had developed yet another problem. The heater switches were not checked and the tank was not replaced, with the result that the next time the tank was filled it effectively became a bomb. That bomb exploded 205,000 miles from Earth.

At T+55hr 55min 20sec, on April 13, a bang was heard in the command module and Jack Swigert saw a warning light come on. "Houston, we've had a problem here," he reported laconically. Capcom Jack Lousma replied: "This is Houston, say again please." At this point James Lovell interjected

sharply: "Houston, we've had a problem. We've had a main B bus undervolt." This piece of shorthand described the alarming symptoms of what had happened, but no-one yet knew that the tank had exploded violently, ripping a 13ft × 6ft panel out of the side of the service module and mortally damaging the tanks and systems inside it.

Within 2hr the crew were without service module oxygen, water, electricity and propulsion. They would have died but for one thing – the lunar module *Aquarius*. Fortunately the astronauts were travelling to the Moon and not back to Earth. The lunar module was still with them and it became their lifeboat. During the tense days that followed, a contingency plan first tried out on Apollo 9 was put into practice. With only 15min of command module power remaining, the crew moved into the lunar module, which was to sustain them until they were within 100 miles of Earth. They fired the descent stage engine for 30sec to put Apollo 13 back on a free-return trajectory. A second burn, lasting 4min 28sec, speeded the

The safe return of Apollo 13 at 12.07 pm Central Pacific Time on April 17, 1970 (Nasa)

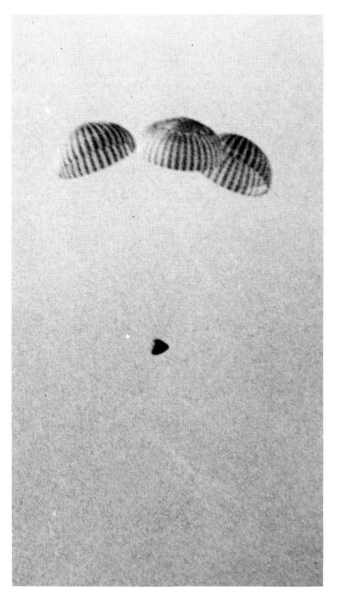

return after the loop around the Moon. Next came a 15.4sec burn to correct a 100-mile mid-course error, and finally a 23sec firing 5hr before splashdown placed the spacecraft in the centre of the re-entry corridor. Throughout the long journey home the crew had huddled in an almost freezing *Aquarius*, conserving the lunar module's stocks of oxygen, water and power, and even jury-rigging a simple air-conditioning unit to absorb their exhaled carbon dioxide.

As Apollo 13 plummeted towards the Earth the damaged service module was discarded by means of a "push-pull" manoeuvre. Some 83min before splashdown the command module *Odyssey* separated from *Aquarius*, which was propelled away by oxygen pressure in the docking collar. *Odyssey* plunged through the atmosphere, with barely enough power to deploy its parachutes. But the great escape was completed and the three astronauts landed safely 3½ miles from USS *Iwo Jima* in the Pacific Ocean at 21°S 165°W.

The ingenuity of thousands of technicians and scientists who made contingency plans and troubleshooting decisions had saved the day, and President Nixon expressed America's relief and gratitude when he said: "Never have so few owed so much to so many." The world's attention was focused once more on Apollo, though it had taken a near-disaster to do it.

Jack Swigert died in December 1982 after a long battle

When Ken Mattingly was dropped from the Apollo 13 crew just days before launch a new crew photo featuring command module pilot Jack Swigert, centre, had to be hastily organised. Fred Haise is on the left and James Lovell on the right (Nasa via Astro Information Service)

against cancer of the bone marrow. He thus became the first "official" astronaut to die of natural causes.

Soyuz 9   June 1, 1970   Flight 52

Name: Soyuz 9
Sequence: 52nd astro-flight, 39th spaceflight, 37th Earth orbit
Launch date: June 1,1970
Launch site: Tyuratam, USSR
Launch vehicle: A2 (SL-4)
Flight type: Earth orbit
Flight time: 17 days 16hr 58min 50sec
Spacecraft weight: About 14,330lb
Crew: Col Andrian Grigoryevich Nikolayev, 40, Soviet Air Force, commander
Vitali Ivanovich Sevastyanov, 34, flight engineer

The last of the pre-space station Soyuz ferry vehicles was Soyuz 9, code-named *Falcon*. It was launched at 10 pm from the Tyuratam Space Centre on June 1, 1970. On board were two men. One was the veteran Andrian Nikolayev, who had flown on Vostok 3 in 1962 and had later married the first spacewoman, Valentina Tereshkova. The couple had a child the following year. Nikolayev graduated from the Zhudovsky Air Force Engineering Academy in 1968, and was back-up commander for the Soyuz 6, 7, and 8 missions.

Vitali Sevastyanov, Soyuz 9's flight engineer, was born on July 8, 1935, in Krasnouralsk in the Sverdlovsk region of the USSR, grew up in Sochi and went to the Ordzhonikidze Aviation Institute. Formerly a lecturer at the Star City cosmonaut training centre, he joined the spacegoing team in 1967. He has written several science papers and broadcast on the radio, and was back-up flight engineer for Soyuz 6, 7, and 8.

The nine-minute A2 launch sequence placed Soyuz 9 in a 128/136-mile orbit with an inclination of 51°. On the seventh

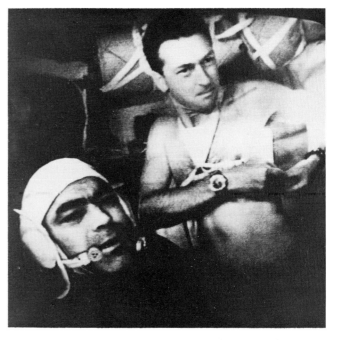

*Nikolayev **left** and Sevastyanov seen during a live telecast from space (Novosti)*

and 17th orbits the Soyuz rocket motor was fired to put the spacecraft into an almost circular path at about 160 miles altitude. Sevastyanov complained that the plume from these burns had smeared his window.

Soyuz 9 at Tyuratam before launch. At the top is the orbital module, in the centre the descent craft and at the base the service module. The solar panels are folded in the launch position (Novosti)

Nikolayev and Sevastyanov in zero-g training (Novosti)

Announcements about the flight indicated clearly that Soyuz 9 was not equipped for rendezvous or docking and that it was to be an extended solo mission on which experiments in the fields of biology, navigation and meteorology would be carried out. Equipment tested included a new type of orientation system based on star-lock navigation, using the bright stars Vega and Canopus in the northern and southern celestial skies. The physical condition of the cosmonauts was monitored carefully, and the relatively spacious orbital compartment contained equipment for extensive daily exercise. This included a torso harness which put a load of about 80lb on the wearer's body when he moved while wearing it. It was hoped that these first serious efforts at exercising in space would mitigate the effects of long-term weightlessness when the crew returned to Earth.

The crew's meteorological and geological observations and photography were compared with simultaneous observations from satellites, aircraft and ships. On the 188th orbit, for example, cloud pictures of the Indian Ocean were taken from Soyuz 9, the Meteor metsat and a ship. The cosmonauts made a number of startling visual sightings, including a tropical storm in the Bay of Bengal on the 69th orbit, forest fires near Lake Chad on the 110th orbit, and the visually elusive luminescent clouds of the upper atmosphere over Arabia on orbit 117.

After nearly 18 days in space, a new world endurance record, Soyuz 9 made a televised descent and a "sniper-accurate" landing in a ploughed field in Intumak, 47 miles west of Karaganda, where it was immediately surrounded by four helicopters. In 424hr 59min Soyuz 9 had covered 7,389,078 miles and completed 286 orbits.

Doctors were eager to see how the cosmonauts were faring after being in space for so long, and the answer was not long in coming. Both men complained of feeling extraordinarily heavy. They had to be carried from the craft and were placed under close medical observation for 10 days. For about a week both crewmen felt that they were being subjected to a force of 2g. They walked with difficulty, becoming flushed and breathing heavily, and stumbled when climbing stairs.

Nikolayev was pessimistic about further long-term flights: "Apparently manned flights of several months will require the development of special measures and means to prepare the organisms of cosmonauts to withstand re-entry g loads and to facilitate re-adaptation to Earth's gravity." Was there a limit to the amount of time man could live in space? The Salyut space stations, their development and launch problems now nearly at an end, were poised to answer that question.

Name: Apollo 14 (AS-509)
Sequence: 53rd astro-flight, 40th spaceflight, 38th Earth orbit, 6th lunar flight, 5th lunar orbit, 3rd lunar landing
Launch date: January 31, 1971
Launch site: Pad 39A, Kennedy Space Centre, Merritt Island, USA
Launch vehicle: Saturn V (SA-509)
Flight type: Lunar landing
Flight time: 9 days 0hr 1min 57sec
Spacecraft weight: 98,137lb (CM/SM/LM)
Crew: Capt Alan Bartlett Shepard Jr, 47, USN, commander
Maj Stuart Allen Roosa, 37, USAF, command module pilot
Cdr Edgar Dean Mitchell, 40, USN lunar module pilot

Though Alan Shepard was the first American in space, most of the laurels went to John Glenn, the first into orbit. It is said that this piqued Shepard, making him doubly keen to fly again and achieve his share of recognition. But before he had a chance to go into space again he was stricken with an inner-ear infection called Menière's disease which caused balance impairment and resulted in his being barred from all types of flying. Bitterly disappointed, Shepard took a desk job as chief of the astronaut corps, but he never gave up hope of flying again. His confidence was justified: by 1969 he was recovered and back on flight status. Much to the chagrin of many in the space team, and with no Apollo or Gemini experience behind him, he was able to use his powerful position to help win selection as the commander of Apollo 14, with two rookies making up the crew. Thus on January 31, 1971, the three men heading for what would be only the third Moon landing had just 15min of space experience between them.

Command module pilot Stuart "Smoky" Roosa was born on August 16, 1933, in Durango, Colorado, joined the Air Force at the age of 20, and gained a degree in aeronautical engineering during his time in the service. He flew F-84F and

Left to right: Edgar Mitchell, Stuart Roosa and Alan Shepard walk down a corridor in the Manned Operations Control Building on their way to the transfer van that will take them to the pad. Behind them is Deke Slayton, the grounded Group 1 astronaut who eventually flew on the Apollo-Soyuz Test Project (Nasa via Astro Information Service)

F-100 fighters, served as a maintenance test pilot, graduated from the Aerospace Research Pilot School at Edwards AFB in 1965 and was an experimental test pilot there until his selection as a Group 5 astronaut in April 1966. He was a member of the support crew for Apollo 9.

The man scheduled to land on the Moon with Shepard was Edgar Mitchell, another member of the April 1966 class of astronauts. Born in Hereford, Texas, on September 17, 1930, Mitchell joined the Navy in 1952. He was amongst the most highly qualified of all the astronauts, with degrees in industrial management and aeronautical engineering and a doctorate in aeronautics and astronautics. He was first in his class at the Air Force Aerospace Research Pilot School, and in 1964/65 he worked as a project manager on the proposed Manned Orbiting Laboratory programme. He was a support crew member for Apollo 9 and back-up lunar module pilot for Apollo 10.

Bad weather delayed the launch for 40min when the count had reached T–8min on January 31, 1971, and it finally took place in very murky weather at 4.03 pm. The transposition and docking manoeuvre during the initial trans-lunar coast was the most problematical yet, running to six docking attempts. There were fears that the docking probe was faulty and that the lunar landing would have to be cancelled because a post-landing docking would not be possible. But, no doubt with the Apollo 13 failure in mind, it was decided to continue the attempt.

Apollo 14 reached lunar orbit and then reduced its pericynthion (nearest point to Moon) to just 10 miles, the nearest a complete Apollo combination ever came to the Moon. Then, just after command module *Kitty Hawk* and

Mitchell walks on the Moon (Nasa via Astro Information Service)

lunar module *Antares* had parted company, a computer fault developed in the LM. The abort system had somehow been activated and nothing would convince the computer that all was in fact well. Shepard and Mitchell had just finished reprogramming their way out of this problem when the landing radar failed. With an abort just seconds away: Mitchell saved the day simply by flicking a switch on and off: "C'mon radar! Get the lock on . . . Phew, that was close." The subsequent descent was incident-free and *Antares* touched down at Fra Mauro with 40sec to spare, only 87ft off target and tilted 6°. The date was February 5, 1971, the time 4.17 am EST.

As Shepard climbed down the ladder, Bruce McCandless, the capcom at Houston, radioed: "Not bad for an old man." Shepard stepped onto the Moon: "It's a long way, but we're here," he said. The fifth man on the Moon was soon joined by the sixth, and the two astronauts set to work deploying the ALSEP and Modular Equipment Transporter (MET). This space-age shopping trolley was to be used to carry tools and Moon rocks – not Shepard when he got tired, as one comic suggested. A static television camera showed spectacular views of the LM and the bulk of Cone Crater 300ft above it, but frustratingly the astronauts did not always remain in view. This first excursion lasted nearly 4hr 49min.

After a sleep the crew left the confines of the lunar module and set off up the slopes of Cone Crater. They had difficulty in judging distances, and while Shepard was convinced they were much further from the rim of the crater than they really were, and were running short of time and should turn back, Mitchell wanted to go on and explore the inside of the crater. "Aw, gee whiz, let's give it a whirl," he said, and an argument ensued. Mitchell, who incidentally was correct, lost and Cone Crater was never explored. "I think you're finks," groused Mitchell. They had actually got to within 150ft of the rim.

Before returning to *Antares* after a walk of 4hr 35min the imperious Shepard broke the ice a little by playing some lunar golf for the sake of the television audience, using his soil-sample scoop as a club. More seriously, he had broken the record for distance travelled on the lunar surface, totalling about 1.7 miles.

Carrying 98lb of Moon rocks, two of which weighed over 10lb each, *Antares* lifted off from Fra Mauro at 1.47 pm EST on February 6 after a stay of 33hr 31min. The earlier docking difficulties were not repeated, and *Kitty Hawk* duly departed lunar orbit after 34 revolutions in 66hr 39min. In that time pilot Roosa had conducted five survey experiments.

Kitty Hawk splashed down on February 9 in the Pacific, four miles from USS *New Orleans* at 27°S 172°W and 765 miles south of Samoa. At $392 million for the mission, each pound of Moon rock had cost four million dollars. Shepard's LEVA time was 9hr 17min, Mitchell's a little less. The flight had taken almost exactly nine days and the distance travelled was 1,157,560 miles.

Shepard works with his "wheelbarrow" at Fra Mauro (Nasa)

Name: Soyuz 10
Sequence: 54th astro-flight, 41st spaceflight, 39th Earth orbit
Launch date: April 23, 1971
Launch site: Tyuratam, USSR
Launch vehicle: A2 (SL-4)
Flight type: Earth orbit, rendezvous and docking
Flight time: 1 day 23hr 45min 54sec
Spacecraft weight: About 14,490lb
Crew: Col Vladimir Alexandrovich Shatalov, 43, Soviet Air Force, commander
Alexei Stanislavovich Yeliseyev, 36, flight engineer
Nikolai Nikolayevich Rukavishnikov, 38, test engineer

On April 19, 1971, soon after the tenth anniversary of the pioneering flight of Yuri Gagarin, the Russians embarked on a slightly delayed celebratory spectacular. Salyut 1, the world's first space station, was launched into orbit.

"Dear comrades and friends, ten years ago our fellow-countryman Yuri Gagarin left on the first spaceflight from this historic cosmodrome. Since then many Soviet cosmonauts have been in near-terrestrial space. Lifting off today, the ship Soyuz 10 will continue the important cause of exploring outer space in the interests of science and the national economy. Our crew will have to carry out an extensive programme of scientific and technical studies and experiments." So said Soyuz 10 commander Col Vladimir Shatalov on the launch pad on April 22 during a build-up as spectacular as his launch with two crewmen at 2.45 am the following day. The three

*With their spacecraft and launcher visible in the background, the Soyuz 10 crew speak to the press. **Left to right** Nikolai Rukavishnikov, Vladimir Shatalov and Alexei Yeliseyev (Novosti)*

cosmonauts had been in intensive training for 14 days, working from 7 am to 11 pm each day. "We prepared so painstakingly," said Shatalov, "you may almost say that the ship is engraved on our eyeballs." Official announcements spoke of a "joint experiment" with Salyut, of a comprehensive check of the "ship," and of medical and biological experiments.

Shatalov and Yeliseyev had flown in Soyuz 4 and 5 respectively, and together in Soyuz 8. Nikolai Rukavishnikov, making his first flight, was born in the Siberian city of Tomsk on September 18, 1932. He entered the Moscow Physics and Engineering Institute in 1951, graduating six years later to become a spacecraft designer. Selected as a cosmonaut in 1967, he was described in Soviet releases as an expert on engineering physics and instrument-making.

Soyuz 10 did suceed in reaching Salyut 1, only to re-enter a few hours later, landing at night in the prime recovery zone 74 miles north-west of Karaganda and narrowly missing a lake and the first Soviet splashdown. Contradictory statements were issued later: the flight had been a total success of the "joint systems"; it had been an "important step"; the docking had imposed a "considerable emotional load on the cosmonauts"; one of the crewmen, Rukavishnikov, had experienced "unusual and unpleasant feelings"; Rukavishnikov had felt "fine". Wherever the truth lies, the flight was an anticlimax and can be classified as only a partial success.

Shatalov, Rukavishnikov and Yeliseyev took a day to reach Salyut 1, which they first sighted when it was nine miles away. An automatic system, with the aid of three orbit changes and four manoeuvres by the supposedly passive Salyut target, brought the craft to a rendezvous 550ft apart. Shatalov then jockeyed Soyuz 10 towards the space station and succeeded in docking, though only after 90min of effort. Was the docking

Soyuz 10 on the pad (Novosti)

too hard, damaging the transfer tunnel? Was Rukavishnikov violently ill? Clearly something had gone wrong, because the cosmonauts did not enter the space station. Soyuz 10

Soyuz 10 lifts off (Novosti)

undocked after 5½hr and apparently made an emergency return to Earth after remaining in orbit for a further 16hr and a total of 32 orbits in nearly two days.

Soyuz 11   June 6, 1971   Flight 55

Name: Soyuz 11
Sequence: 55th astro-flight, 42nd spaceflight, 40th Earth orbit
Launch date: June 6, 1971
Launch site: Tyuratam, USSR
Launch vehicle: A2 (SL-4)
Flight type: Ferry to Earth-orbital space station
Flight time: 23 days 18hr 21min 43sec
Spacecraft weight: About 14,475lb
Crew: Lt-Col Georgi Timofeyevich Dobrovolsky, 43, Soviet Air Force, commander
Vladislav Nikolayevich Volkov, 35, flight engineer
Viktor Ivanovich Patsayev, 38, test engineer

The Soyuz 10 crew had arrived at their Salyut 1 space hotel to find the door locked. It was left to Georgi Dobrovolsky, Vladislav Volkov and Viktor Patsayev to check in for an extended stay two months later.

Dobrovolsky was born on June 1, 1928, in Odessa. He attended air force school in Chuguyeve and graduated from the Air Force Academy in 1961. Two years later he joined the cosmonauts. Patsayev came from Aktyubinsk, where he was born on June 19, 1933. He graduated from the Penze Industrial Institute with a master's degree and worked as a radio researcher and design engineer at the Central Aerological Observatory. He became a cosmonaut in 1969. Volkov had flown on Soyuz 7.

Using the callsigns *Yantar* 1, 2 and 3, the cosmonauts of Soyuz 11 launched at 7.25 am on June 6 and made an automatic approach to within 100m of the space station. Dobrovolsky performed a manual docking and within hours the crew were settling in as the first men to occupy a space station. The combined Soyuz/Salyut "orbital complex," as the Russians called it, weighed 55,000lb (less than Apollo), was 66ft long and contained 3,500ft³ of workspace.

Salyut 1, launched by the 2,400 000lb-thrust D1 (SL-13) on April 19, was modular in construction. The 6.5ft-diameter

Inside Soyuz 11 are Dobrovolsky **centre**, Patsayev **left** and Volkov
(Novosti)

transit compartment incorporated the Soyuz docking collar, contained communications equipment and a celestial telescope. Next to this was the 13ft-diameter main work station and sleeping/rest quarters. At the rear were the unpressurised engine propellant and ancillary equipment bays. Attached to the exterior of the station were four solar panels in two pairs.

Dobrovolsky, Volkov and Patsayev soon set to work on a highly successful three-week scientific and biomedical programme. They studied the stars using an optical and spectrographic telescope designated Orion. Weather reports were made and observations compared with Meteor satellite photographs. Earth-resources photography of the land and sea reaped a harvest of data. The cosmonauts grew crops, including flax, in their Oasis greenhouse and hatched frogs' eggs and carried out genetic studies of fruit flies. The atmosphere of the Earth was studied in detail, and the cosmonauts noted a high-frequency resonance on the Salyut transmitting antennae when the craft passed through low-temperature plasma. "Penguin" elastic restraining suits were worn for exercise purposes and intensive biomedical work was conducted.

In all, the Soyuz 11/Salyut 1 mission had the makings of one of the most successful and scientifically rewarding in history. As Salyut 1 made its 1,000th orbit at 2.14 am Moscow time on June 20, the cosmonauts had completed 206 orbits and Patsayev was celebrating his 39th birthday. Nine days later the crew were transferring to Soyuz 11 and powering up its systems for their return to Earth. Wearing the now customary woollen flight suits and leather helmets rather than pressurised spacesuits, the cosmonauts settled down in their couches and separated their craft from Salyut 1. They briefly flew in formation with the space station and then fired the retro-rockets. The onset of very slight g forces after about 10min signalled the start of re-entry, and at that point the orbital compartment was separated from the flight cabin by the firing of explosive bolts. It was then that the textbook mission started to go terribly wrong.

The explosive separation sequence shook open an exhaust valve in the flight cabin. The resulting jet of air from the cabin disorientated the craft, but this was corrected by the

This fine picture of a Soyuz launch reveals the full extent of the A2 and the tapered lower stage and strap-on boosters (Novosti)

automatic control thrusters. These unexpected gyrations confused the crew, however, delaying their response to the open valve. It took just 45sec for the capsule to depressurise completely, by which time the cosmonauts were probably already dead. The disaster was not discovered until the remainder of the re-entry, carried out under automatic control, had been completed about 20min later. The capsule landed on target about 500 miles south-west of Sverdlovsk, where recovery crews found the cosmonauts "as if sleeping".

Twelve days later the Russians announced the cause of death as "depressurisation," but it was not until three years later, during preparations for the ASTP joint US-USSR flight, that they finally explained the exact reason. Dobrovolsky, Volkov and Patsayev lay in state and were buried with full state honours in Moscow. American astronaut Tom Stafford attended.

If the crew had worn spacesuits they would have survived the emergency. On Voskhod 1 and Soyuz 3, 4, 5, 6, 7, 8, 9 and 10 the cosmonauts had done without this fundamental form of protection, all in the cause of cramming an extra man aboard. Soyuz 11 was one cut corner too many, and the three cosmonauts paid the penalty.

Salyut 1 was never inhabited again and was commanded to re-enter the Earth's atmosphere on October 11, 1971, after 2,800 orbits when it became clear that the space station would soon run out of the propellent need to sustain its low orbit. The Soyuz redesign made necessary by the accident put a stop to the Soviet manned spaceflight programme for the next two years.

An artist's impression of Soyuz 11 docking with Salyut 1 (Novosti)

Apollo 15   July 26, 1971   Flight 56

Name: Apollo 15 (AS-510)
Sequence: 56th astro-flight, 43rd spaceflight, 41st Earth orbit, 7th lunar flight, 6th lunar orbit, 4th lunar landing
Launch date: July 26, 1971
Launch site: Pad 39A, Kennedy Space Centre, Merritt Island, USA
Launch Vehicle: Saturn V (SA-510)
Flight type: Lunar landing
Flight time: 12 days 7hr 11min 53sec
Spacecraft weight: 103,143lb (CM/SM/LM)
Crew: Col David Randolph Scott, 39, USAF, commander
Maj Alfred Merrill Worden, 39, USAF, command module pilot
Lt Col James Benson Irwin, 41, USAF, lunar module pilot

The Apollo 15 flight seemed to have everything: a spectacular Moon-landing site, the first lunar rover, continuous colour television coverage of the highest quality, and a lift-off to remember.

On launch day the three astronauts were glimpsed briefly by spectators as they left the Manned Spacecraft Operations building and boarded the transfer van which took them to the launch complex. First out was David Scott, followed by Irwin and Worden, who stopped to say goodbye to members of his family. The crew's departure took place in near silence apart from a few shouts, cheers and ripples of applause, the watching pressmen being too busy taking photographs.

The launch took place at 9.34 am on a very hot, steamy day. Seen immediately beforehand from the press site three miles away, the Saturn V launcher sat motionless on the pad, serenely puffing clouds of vaporising liquid oxygen fuel. Then, with the launch imminent, silence suddenly fell over the press site. By the base of the great rocket a ball of rich orange flame blossomed, and steam billowed out to left and right of the pad. Ponderously the Saturn rose, poised on a tongue of flame that licked the launch complex and tower. Then came the noise, heralded by shock waves that rippled the waters of the Cape's lagoons. For two minutes onlookers were subjected to a shattering roar and crackle that shook the ground and hammered at bodies and eardrums. Minutes later, high in the clear blue sky, the launch escape system and first stage could be seen falling away as Apollo 15 headed smoothly for Earth orbit and the trans-lunar coast.

David Scott was on his third flight, having served as Apollo 12 back-up commander after his Apollo 9 mission. Command module pilot Alfred Worden came from Jackson, Michigan, where he was born on February 7, 1932. He graduated from the military academy in 1955, joined the Air Force and later graduated from the Empire Test Pilots School in Farnborough, England. Worden then entered the Aerospace Research Pilot School at Edwards AFB and subsequently became an instructor there. He joined the fifth astronaut group in 1966 and served as a support crew member for Apollo 9 and back-up CMP for Apollo 12.

The back-up LMP for Apollo 12 was James Irwin, who thus found himself in line for Apollo 15. Lucky to survive a serious jet crash at Edwards in 1961, Irwin joined the 1966 astronauts and served as an Apollo 10 support crew member. He was also the crew commander for the LTA-8 lunar module thermal vacuum test simulation in 1968. Irwin was born on March 17, 1930, in Pittsburgh and graduated in naval sciences, aeronautical engineering and instrumentation engineering before joining the Air Force from the Naval Academy in 1951. He graduated from the Aerospace Research Pilot School in 1963.

Another space first-timer on the flight was the lunar roving vehicle. Measuring 10ft long by 6ft 9in wide, it had four individually powered wire-mesh wheels and weighed 1,540lb when loaded. The LRV was ingeniously packed into the side of the lunar module and unfolded on the Moon at the pull of a few lanyards. It was powered by two silver-zinc batteries and had a top speed of 8.7mph.

A steep, 26° descent over the Appenine Mountains was followed by a safe landing at Hadley Base by lunar module *Falcon* at 11.16 pm BST on July 30. Scott and Irwin were soon taking the Earth television audience for a spectacular full-colour tour of the Hadley Rille and St George's Crater with the aid of a camera and antenna mounted on the LRV.

The astronauts worked hard, fell a few times and collected some interesting samples, one of which Scott called the "Genesis Rock" because he assumed it, mistakenly as it turned out, to be pristine lunar material. The urbane commander also carried out Galileo's proposed demonstration of the effect of gravity in vacuum. He dropped a hammer and a falcon's feather simultaneously and the two objects hit the ground at the same time. "Galileo was right," chortled Scott. The three EVAs lasted a total of 18hr 37min, and the astronauts travelled 17.3 miles in the LRV. They also deployed an ALSEP array of instruments, collected a core sample from a depth of 10ft, and accumulated a total of 173lb of Moon rock.

After 66hr 55min on the Moon *Falcon*, tilted at 10°, separated from the descent stage. The lift-off was watched

Apollo 15 crew at the Vertical Assembly Building to witness the roll-out of their Apollo-Saturn vehicle. **Left to right** *James Irwin, Alfred Worden and David Scott (Nasa)*

An eve-of-launch view of Apollo 15 (Author)

James Irwin salutes the flag at Hadley Base. On the right is the Lunar Rover and in the background are the Appennine Mountains. This was the most spectacular of the six sites visited by the Apollo astronauts. (Nasa)

live by an Earth audience thanks to the camera on the Rover, and accompanied by the strains of *The Wild Blue Yonder*, being played by Worden in command module *Endeavour*. Scott and Irwin had clocked up individual LEVA times of 17hr

36min and 17hr 11min respectively. Scott had also performed a 35min SEVA soon after landing, putting his head out of the top of the LM for a preliminary look round. Meanwhile, Worden had been extremely busy operating 16 science experiments and conducting a detailed lunar survey using equipment housed in the Scientific Instrument Module bay (SIMBAY) in the service module.

Endeavour ejected the first ever Apollo subsatellite in lunar orbit and then headed for Earth with a 2min 21sec SPS burn after 74 orbits of the Moon in six days 1hr 18min. During the return flight Worden made a planned deep-space EVA 199,000 miles from Earth, the first of its kind, to retrieve instrument packages from the side of the service module. The 38min EVA was televised, and during it Irwin stood in the hatchway of the command module. What the viewers did not see and what Worden did not have time to photograph was what he called "an absolutely fantastic sight": Irwin standing against the background of a full Moon in the darkness of space.

The 1,282,256-mile journey of Apollo 15 ended on August 7 with a splashdown five miles from USS *Okinawa* and 335 miles north of Honolulu at T+295hr. The last phase of the recovery had been marred by one of the few flaws of the mission: one of the three parachutes failed following contamination during a fuel dump. The command module landed at 21mph instead of 19mph but the astronauts were unhurt.

Scott, Irwin and Worden returned from the most productive lunar mission yet to find themselves in deep trouble. For franking envelopes on the Moon and selling them on behalf of a trust fund for their children, and because duplicates of statuettes they had taken with them found their way on to the market, the three astronauts were reprimanded and dropped from flight status.

Apollo 16   April 16, 1972   Flight 57

Left to right: John Young, Charlie Duke and Ken Mattingly after their recovery from the Apollo 16 command module in the Pacific Ocean (Nasa via Astro Information Service)

Name: Apollo 16 (AS-511)
Sequence: 57th astro-flight, 44th spaceflight, 42nd Earth orbit, 8th lunar flight, 7th lunar orbit, 5th lunar landing
Launch date: April 16, 1972
Launch site: Pad 39A, Kennedy Space Centre, Merritt Island, USA
Launch vehicle: Saturn V (SA-511)
Flight type: Lunar landing
Flight time: 11 days 1hr 51min 5sec
Spacecraft weight: 103,165lb
Crew: Capt John Watts Young Jr, 41, USN, commander
Lt-Cdr Thomas Kenneth Mattingly, 36, USN, command module pilot
Lt-Col Charles Moss Duke Jr, 36, USAF, lunar module pilot

By the time Apollo 16 came to be launched, at 12.54 pm on April 16, 1972, Apollos 18, 19 and 20 had all been cancelled and everyone knew that this would probably be the 20th century's penultimate voyage to the Moon. Public interest was dwindling and Apollo 16 was in danger of becoming anonymous. But the Jeremiahs hadn't reckoned on lunar

John Young, clad in heavily dust-soiled lunar surface suit and carrying a geological sample bag, works at North Ray Crater (Nasa via Astro Information Service)

Charles Duke walks the lonely Descartes plains, leaving footsteps which could remain undisturbed for millions of years (Nasa)

module pilot Charles Duke, whose infectious enthusiasm, humour and Carolina drawl endeared him to the public even though it might have annoyed those scientists who didn't believe that working on the Moon should be fun.

Duke was born in Charlotte, North Carolina, on October 3, 1935, and graduated brilliantly from the Naval Academy, earning degrees in naval sciences and in aeronautics and astronautics. After joining the Air Force in 1957 he graduated with distinction from training and entered the Aerospace Research Pilot School, where he was serving as an instructor when chosen as an astronaut in 1966. Duke was on the Apollo 10 support crew and was specifically requested by Armstrong to be capcom for the Apollo 11 landing phase. He was also Apollo 13 back-up LMP: It was his catching German measles that led to the grounding from that flight of Thomas Mattingly, who was rightfully given the CMP job on Apollo 16.

Mattingly was born in Chicago on March 17, 1936. He gained an aeronautical degree in 1958 and joined the Navy, receiving his wings in 1960. When he joined the fifth class of Nasa astronauts he was a student at the Aerospace Research Pilot School. Mattingly served on the support crews for Apollos 8 and 11.

Mission commander John Young was making his fourth spaceflight and his second into lunar orbit, and would be the ninth man to set foot on the Moon. Like his crew, he had served as a back-up on Apollo 13.

Apart from a few minor faults the flight proceeded well until the LM and CM separated in lunar orbit and became *Orion* and *Caspar*. Then, as Mattingly prepared to fire the SPS engine for a circularisation burn, he found a fault in the yaw gimbal drive servo loop. The landing was delayed and while Young and Duke sweated it out in *Orion*, Mattingly in *Caspar* tried to solve a problem that could have jeopardised the astronauts' ability to leave lunar orbit. With only two orbits left before the landing would have to be cancelled because of the increasing inaccessibility of the site from the LM's orbit, Mattingly finally corrected the fault. Young and Duke landed at Descartes Base in the highlands of the Moon 5hr 43min

late. Duke's enthusiastic commentary went like this: "Coming in like gangbusters. I can see the landing site. We're right in, John. Man, there it is. Gator. Lone Star. Perfect place over there, John. A couple of big boulders – contact – stop. Well, *Orion* is finally here, Houston. Fantastic! All we have to do is jump out of the hatch and we've go plenty of rocks . . . Man, it really looks nice out there . . . I'm like a little kid on Christmas Eve."

Later Young stepped down from the ladder of the LM and said: "Here you are, mysterious and unknown Descartes, highland plains. Apollo 16 is going to change your image." Joined by Duke, he went about the serious business of setting out an array of ALSEP scientific instruments, including an astronomical observatory, and deploying the LRV. Unfortunately, Young tripped over a wire and ruined a particularly interesting heat-flow experiment. Duke was most upset, having spent a lot of time setting it up. Much of the three EVAs was taken up by drives in the LRV, totalling 16.8 miles, to survey as much of the area as possible and to collect a fine bag of rocks. Once again television viewers were treated to spectacular vistas of the Moon and the sight of cavorting, laughing astronauts. Duke and Young even tried at the end of their explorations to perform a "high jump" experiment. The lively Duke tried especially hard and fell over, landing on his vital backpack. It could have been disastrous but no harm was done. Young also drove the LRV at high speed, skidding it repeatedly to assess the degree of wheel grip.

During a rest period in the LM the astronauts inadvertently left their microphones switched on and the world was treated to some explicit descriptions of a digestive problem they were experiencing as a result of drinking a lot of potassium-laced orange juice to help their heart function.

The astronauts collected 213lb of Moon rock from Descartes which at 18,000ft above lunar "sea level" was the highest site explored on Apollo. Their departure from the site at 8° 59'S 15° 30'E was recorded by a camera on the lunar

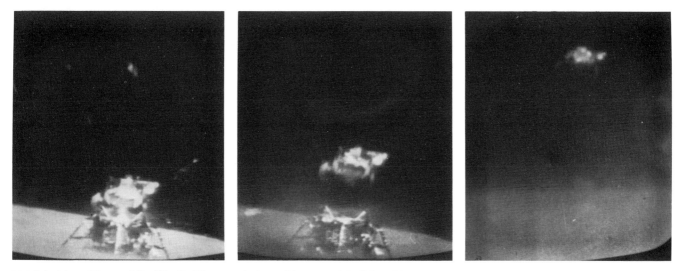

Live television pictures of the lift-off of the ascent stage of lunar module Orion (Nasa)

rover and the pictures shown live on Earth. Young and Duke had been on the Moon for 71hr 14min and the EVAs had lasted 7hr 11min, 7hr 23min and 5hr 40min, all timed from depressurisation to repressurisation of the LM. After a safe docking with *Caspar* the crew jettisoned the ascent stage of the LM, which became unstable and briefly threatened to collide with the CM before the crew took evasive action. Because the SPS problem had not been totally solved the lunar orbit phase of the mission ended one day early. Even so, *Caspar* had flown around the Moon for 125hr 53min and 64 orbits, and Mattingly had flown the longest US solo

spaceflight (81hr 40min) and conducted 15 SIMBAY experiments. His last task in the neighbourhood of the Moon was to eject a subsatellite, while on the way home he made the now customary deep-space EVA, lasting 1hr 24min.

Television viewers saw parachute deployment and splashdown less than a mile from USS *Ticonderoga* in the Pacific at 0° S 156° W. This was the most accurate manned landing to date, closer even than Gemini 9. The astronauts reached the carrier deck in 37min, the shortest Apollo recovery time. Apollo 16 travelled 1,399,377 miles in 11 days. There was only one more Moon flight to come.

Apollo 17 December 7, 1972 Flight 58

Name: Apollo 17 (AS-512)
Sequence: 58th astro-flight, 45th spaceflight, 43rd Earth orbit, 9th lunar flight, 8th lunar orbit, 6th lunar landing
Launch date: December 7, 1972
Launch site: Pad 39A, Kennedy Space Centre, Merritt Island, USA
Launch vehicle: Saturn V (SA-512)
Flight type: Lunar landing
Flight time: 12 days 13hr 51min 59sec
Spacecraft weight: 103,187lb (CM/SM/LM)
Crew: Capt Eugene Andrew Cernan, 38, USN, commander
Cdr Ronald Ellwin Evans, 39, USN, command module pilot
Harrison Hagen "Jack" Schmitt, 37, lunar module pilot

When Joe Engle was made back-up lunar module pilot for Apollo 14 in 1969 he could have been forgiven for thinking ahead eagerly to the day when he would walk on the Moon during the Apollo 17 mission. Unfortunately for Engle, Apollos 18, 19 and 20 were cancelled and geologist-astronaut Jack Schmitt, scheduled to fly on Apollo 18, was drafted in his place. This was the last chance to put a trained geologist on to the Moon, and for once the scientific lobby got its own way. As the only geologist in the astronaut corps,

Schmitt was therefore given the very last Moon-landing berth in the Apollo programme.

Schmitt was born on July 3, 1935, in Santa Rita, New Mexico, and won a science degree in 1957 and a doctorate in geology from Harvard in 1964. Before becoming one of the

The Apollo 17 crew: **left to right** *Harrison Schmitt, Eugene Cernan and Ronald Evans* (Nasa)

The cataclysmic moment of blast-off as Apollo 17 prepares to turn night into day (Nasa via Astro Information Service)

Earthrise from Apollo 17 (Nasa via Astro Information Service)

first scientist-astronauts in 1965 he was project chief for lunar field geological methods at the US Geological Survey's Astro-geology Branch in Flagstaff, Arizona, and instructed astronauts during field trips. Following his selection as an astronaut Schmitt had to learn how to fly fast jets. He served as Apollo 15 back-up LMP before being selected for Apollo 18 and then Apollo 17.

Apollo 17 CMP was Ronald Evans, one of the 1966 class of Nasa astronauts. Evans was born on November 10, 1933, in St Francis, Kansas. He joined the US Navy in 1957, gained degrees in electrical and aeronautical engineering, and was one of the few astronauts to have served in the Vietnam War,

flying F-8 fighters from USS *Ticonderoga*. He served as back-up CMP for Apollo 14 and on the support crews for Apollo 7 and 11.

Eugene Cernan, now a grey-haired 38-year-old, was in command of Apollo 17, having served as Apollo 14 back-up commander. Cernan called the CM *America* and the LM, in which he and Schmitt would land in a valley at Taurus Littrow, *Challenger*.

The final flight to the Moon was to begin with the first night launch of a US manned spacecraft and of a Saturn V. It promised to be spectacular and it was, turning midnight (00.33 am) into day for miles around. But spectators had had

Last footprints on the Moon at Taurus Littrow as Cernan poses for Schmitt, seen reflected in his commander's helmet (Nasa via Astro Information Service)

Ronald Evans pictured during his trans-Earth EVA (Nasa)

to wait an extra 2½hr to see it because at T–30sec the launch was delayed when a computer refused to accept a manual correction to a pressurisation fault in the third stage.

Challenger landed on the Moon at 20° 10′N 30° 45′E with 2min of fuel left. Schmitt expressed the opinion that Cernan might have hovered a while to select the best possible landing site. "I like it right where we are," Cernan replied. The subsequent television pictures of Taurus Littrow, the mountains, the huge boulders, the hard-working and dusty astronauts, lunar rover and the LM nestled in a little crater were quite superb. Viewers could even see Cernan's face when for a short while he lifted the outer visor of his helmet. But the high point of the mission came when Schmitt spotted some orange soil which the astronauts thought, wrongly as it turned out, could have indicated the presence at one time of water on the Moon.

Challenger remained on the Moon for 74hr 59min. The EVAs lasted 7hr 12min, 7hr 37min and 7hr 16min. The astronauts collected 243lb of Moon samples and deployed an array of ALSEP instruments. They had also driven a total of 21 miles in the LRV. Cernan was the last Apollo astronaut to leave the Moon, and before he stepped on to the LM ladder he said: "We leave as we came, in peace for all mankind."

Ronald Evans in *America* had meanwhile been making detailed surveys of the surface from orbit. When the lunar walkers returned and *Challenger* had been ejected to impact 10 miles from the landing site, the command module ejected a subsatellite and bade farewell to the Moon after 147hr 48min and 75 revolutions in orbit. Evans made a televised deep-space EVA lasting 1hr 6min on the way home, and the 12-day mission ended with a splashdown just 3.9 miles from USS *Ticonderoga* in the Pacific.

The Apollo programme was at times bitterly criticised on account of its political origins and immense, $24,000 million, cost. But the fact remains that it achieved a great deal for science. Twelve men had spent 160 man-hours on the Moon, covering 60 miles on foot and by LRV and collecting 2,196 samples of rock weighing a total of 850lb and varying in age between 3,100 and 4,700 million years old. The samples, divided into over 39,000 pieces, kept scientists in 19 countries all over the world very busy for many years. More than 30,000 photographs of the Moon were taken, and 60 major experiments were placed on the surface and 30 carried out from lunar orbit.

Skylab 2   May 25, 1973   Flight 59

Name: Skylab 2
Sequence: 59th astro-flight, 46th spaceflight, 44th Earth orbit
Launch date: May 25, 1973
Launch site: Pad 39B, Kennedy Space Centre, Merritt Island, USA
Launch vehicle: Saturn IB (SA-206)
Flight type: Ferry to Earth-orbital space station
Flight time: 28 days 0hr 49min 49sec
Spacecraft weight: 30,384lb
Crew: Capt Charles "Pete" Conrad Jr, 42, USN, commander
Cdr Joseph Peter Kerwin, 41, USN, science pilot
Cdr Paul Joseph Weitz, 40, USN, pilot

Skylab was America's first – and so far only – space station. It was born in 1966 as the Apollo Applications Programme, designed to utilise Apollo hardware not used in lunar landings and to ensure an American presence in space after those landings. Basically a converted Saturn S-IVB stage, Skylab 1 comprised the cylindrical orbital workshop, to which were attached two large solar panels, an airlock module, a multiple docking adapter with two docking ports, and the Apollo Telescope Mount, a converted lunar module with four extendable solar panels. The space station, with its working volume of 13,000ft³, was to house three teams of three astronauts for 28, 56 and 56 days respectively. During that time they would conduct the most comprehensive scientific experiment programme undertaken in space, carrying out a planned 270 life sciences, solar physics, Earth-observation, astrophysics, materials-processing, engineering and technology, and student experiments.

But all this seemed out of the question on May 14, 1973, when Skylab appeared to be a useless orbiting hulk, crippled by a launch accident and later malfunctions. All that could be said for the hapless Skylab was that at 164,896lb it was the heaviest object ever placed in orbit.

The final Saturn V, AS-513, left Pad 39A on May 14. Just 63sec later, with the launcher climbing through thick cloud, the Skylab micrometeorite thermal shield tore loose, taking with it one of the solar wings. To make matters even worse, the second wing, jammed by debris, failed to deploy in orbit. The problem was discovered 41min into the mission and Skylab seemed doomed, drifting uselessly without power or thermal protection.

Joe Kerwin, left, Pete Conrad and Paul Weitz, the all-Navy crew of Skylab 2, pictured in the Skylab simulator (Nasa via Astro Information Service)

Conrad, background, and Kerwin seen during the daring but successful EVA that saved Skylab (Nasa)

The first crew to inhabit Skylab were to be launched the following day but had to wait until May 25 for repair methods to be developed and tested. This effort was so intensive that a vital piece of equipment, a parasol designed to reduce the temperature inside Skylab, was crammed into the Apollo command module ferry vehicle only hours before the launch from Pad 39B. Skylab 2's crew had quite a job on their hands, and their mission was humorously summarised by a cartoonist who drew a screwdriver blasting off.

Though mission commander Charles Conrad was no newcomer to space, his two colleagues were. The science pilot was Cdr Joe Kerwin, chosen as a scientist-astronaut in 1965. He was born on February 19, 1932, in Oak Hill, Illinois, and gained degrees in philosophy and medicine before becoming an intern at Washington DC's general hospital. Kerwin then joined the Naval School of Aviation Medicine and won his pilot's wings in 1962. When selected as an

The Skylab 2 crew leave the space station in a good state of repair. A sunshade has been erected and the remaining solar panel has been extended (Nasa)

astronaut he was serving as flight surgeon on a fighter squadron.

The pilot was Paul Weitz, also a Navy man. He was born on July 25, 1932, in Eire, Pennsylvania, and had two degrees in aeronautical engineering. Weitz joined the Navy in 1954 and flew combat missions in Vietnam before becoming an astronaut in 1966. He served on the Apollo 12 support crew.

When Conrad reached Skylab he couldn't believe his eyes. One solar array had been torn off, he reported, and one was stuck in the closed position. The astronauts opened the hatch of the command module, Conrad flew in close to the space station and Weitz, using a pole with a hook at the end, tried to pull the array open while Kerwin hung on to his spacesuited ankles. Weitz persevered for an hour but to no avail. The astronauts then decided to dock, but it took 2hr and no fewer than eight attempts. Skylab seemed to be a jinxed mission.

The crew later entered the space station, now 118ft long with the addition of the CM/SM, and erected the parasol on the exterior of the workshop. By the fourth day the interior had cooled to a comfortable level, and on the seventh Conrad and Kerwin performed a hazardous EVA during which, using wire cutters, Conrad freed the jammed wing and deployed it. Skylab had power and the project was saved.

Conrad, Kerwin and Weitz worked on in Skylab and came home after 404 revolutions to USS *Ticonderoga*, waiting to the south-west of San Diego. The 28-day mission had been the longest in history and the crew were remarkably fit and well. They had managed to complete 46 of the planned 55 experiments, clocking up 392hr of experimental time, and each had gained EVA experience: Conrad 4hr 14min, Kerwin 3hr 30min and Weitz 2hr 21min. The first space repairmen had travelled nearly 14 million miles and cleared the way for two more richly productive Skylab missions.

Skylab 3   July 28, 1973   Flight 60

Name: Skylab 3
Sequence: 60th astro-flight, 47th spaceflight, 45th Earth orbit
Launch date: July 28, 1973
Launch site: Pad 39B, Kennedy Space Centre, Merritt Island, USA
Launch vehicle: Saturn IB (SA-207)
Flight type: Ferry to Earth-orbital space station
Flight time: 59 days 11hr 9min 4sec
Spacecraft weight: 30,561lb
Crew: Capt Alan Lavern Bean, 41, USN, commander
Owen Kay Garriott, 42, science pilot
Maj Jack Robert Lousma, 37, USMC, pilot

Because the problems on Skylab seemed likely to reduce its operational life, Skylab 3 was pressed into service earlier than anticipated. Men, mice, fish, gnats and spiders were inside the command module at Pad 39B when the Saturn IB lifted off

Left to right: Owen Garriott, Alan Bean and Jack Lousma prepare for Skylab 3 in the Orbital Workshop trainer (Nasa via Astro Information Service)

at 7.11 am on July 28. The spiders were called Anita and Arabella, the men Bean, Garriott and Lousma.

Bean's last assignment was as lunar module pilot on Apollo 12, making him the fourth man on the Moon. Owen Garriott, the science pilot, was the third of the scientist-astronauts, chosen in 1965, to fly. There were originally six but two resigned, conveniently leaving Schmitt for Apollo 17 and three (Kerwin, Garriott and Gibson) for Skylab. Dr Garriott was an electrical engineer, having gained bachelor's and master's degrees and a doctorate in 1953, 1957 and 1960 respectively. Before joining the astronaut corps he had taught electronics, electromagnetic theory and ionospheric physics at Stanford University. Garriott was born on November 22, 1930, in Enid, Oklahoma.

Jack Lousma was one of the few Marine astronauts, and only the second to fly. Born in Grand Rapids, Michigan, on February 29, 1936, he earned degrees in aeronautical engineering and joined the Marines in 1959, receiving his wings a year later. A member of the 1966 astronaut class, Lousma served on the support crews for Apollos 9 and 13 and was the capcom recipient of the famous "Houston, we've had a problem" message from Apollo 13.

The astronauts suffered from motion sickness when they arrived at Skylab, delaying the start of work, and later had further problems with a leaking thruster quad on the service module of their ferry vehicle. Then another quad showed similar symptons, presenting the possibility that the astronauts were stranded in space with a crippled ferry vehicle. For the first time ever Nasa was in a position to attempt a space rescue: a back-up Apollo with five seats and two pilots, Vance Brand and Don Lind, would be launched, to dock at the second port and bring the crew home. But the problem was not as serious as first thought, and Brand and Lind's unusual first flight was cancelled.

The Skylab 3 crew performed well during a record stay in the space station. Bean tested the astronaut manoeuvring unit developed for Gemini, and all three made spacewalks to retrieve photo cassettes and samples. They deployed a new twin-pole parasol, replaced six rate gyros, repaired nine pieces of equipment, and handsomely improved on the science activity plan. Many mission goals were exceeded by

50 per cent, 305 man-hours out of the total of 1,081hr of experimentation being spent on solar observations alone. Some 16,000 photographs and 18 miles of Earth-resources data tape were collected, and the astronauts conducted 333 medical experiments. Bean clocked up 12hr 45min of EVA time, Lousma 10hr 59min and Garriot 13hr 44min. Anita and Arabella spun webs in zero g, proving the creatures' rapid adaptability. Anita died in orbit and Arabella at the Marshall SFC after the flight.

On September 25, 1973, the three returned to USS *New Orleans* in the Pacific south-west of San Diego, having survived with apparent ease a record 59 days of weight-lessness. The three had travelled around the Earth 892 times and covered 24,400,000 miles.

Of science pilot Garriott's performance Bean had this to say: "The flight was 50 per cent more productive having Owen aboard. I'm really sold on these scientist-astronauts." This no doubt brought a wry smile to the faces of the "Excess 11" scientist-astronauts, chosen in 1967 but by now with little hope of ever flying.

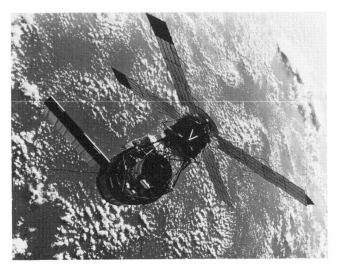

A pre-docking view of the space station (Nasa)

Lousma pictured during his EVA with Owen Garriott (Nasa via Astro Information Service)

84

Name: Soyuz 12
Sequence: 61st astro-flight, 48th spaceflight, 46th Earth orbit
Launch date: September 27, 1973
Launch site: Tyuratam, USSR
Launch vehicle: A2 (SL-4)
Flight type: Earth orbit
Flight time: 1 day 23hr 15min 32sec
Spacecraft weight: About 14,450lb
Crew: Lt-Col Vasili Grigoryevich Lazarev, 45, Soviet Air Force, commander
Oleg Grigoryevich Makarov, 40, flight engineer

Soyuz 12's Oleg Makarov **left** *and Vasili Lazarev* (Novosti)

The Soyuz 12 crew in a ground trainer (Novosti)

The Soyuz 11 disaster did not mark the end of the Soviet run of space mishaps. Salyut 2 broke up in orbit and Cosmos 557, launched at about the same time as the American Skylab, was also a failed Salyut. Designed to fly to this space station was a new type of Soyuz, incorporating extra safety measures. Chief among these changes was the deletion of the third crew position so that the remaining crew could fly fully space-suited. The craft was powered by batteries rather than solar panels, the resulting two-day independent endurance in space being regarded as sufficient for rendezvous and docking with a space station.

Because the Cosmos-Salyut failed the Russians probably decided to test-fly the new Soyuz on a solo, independent mission. When Soyuz 12 was launched at 3.18 pm on September 27 they uncharacteristically announced that the craft would remain in space for just two days, probably to avoid rumours of yet another failure after an apparently truncated mission. Not that a failure seemed likely, the new Soyuz having been tested successfully on a number of Cosmos flights. The crew were not experienced in space but were veterans all the same. The commander, Vasili Lazarev, was 45. Born on February 23, 1928, in the Altai region of southern Siberia, he trained as a doctor before becoming a pilot. He applied without success to join the first group of cosmonauts in 1960, but was later drafted into the team to act as back-up Voskhod 1 doctor after becoming a cosmonaut in 1964. He was back-up commander of Soyuz 9. Back-up flight scientist for Voskhod 1 was Oleg Makarov, who joined Lazarev for Soyuz 12. Makarov was born on January 16, 1933, in the Kalinin region near Moscow and went to a technical college in 1951. He became a designer in 1957 in the department that built Sputnik 1 and joined the scientist-cosmonauts in 1964.

Soyuz 12 was essentially a test of the ferry craft's manoeuvrability, and little else was done on the flight apart from some Earth-resources photography. The spacecraft went into a high, almost circular 200-mile orbit and carried out extensive manoeuvres to simulate rendezvous. The re-entry sequence was the same as in earlier Soyuz missions, with the orbital compartment, equipped with the docking probe, being ejected first. There were probably some high cosmonaut heart rates at this point: the last time this manoeuvre had been carried out, over two years earlier on Soyuz 11, three cosmonauts had died.

The re-entry module descended to 27,000ft at a rate of about 700ft/sec before a barometric system triggered the automatic parachute-deployment sequence. The stowage cover was jettisoned and small pilot chutes were ejected to drag out the braking parachute, which slowed the descent rate to about 270ft/sec. The main parachute was then deployed, though only partially at first, reducing the speed to about 100ft/sec. When the chute was fully deployed the speed was reduced still further, to about 18ft/sec. The heatshield was jettisoned, exposing deceleration rocket nozzles and switching on a radio recovery beacon. At about 6ft above the ground the rockets fired, easing the craft down to a soft landing. A reserve parachute system would have ensured a landing at a survivable 30ft/sec. The successful test of the Soyuz ferry ended at 2.34 pm Moscow time, two days and 32 orbits after launch, and about 248 miles south-west of Karaganda.

Name: Skylab 4
Sequence: 62nd astro-flight, 49th spaceflight, 47th Earth orbit
Launch date: November 16, 1973
Launch site: Pad 39B, Kennedy Space Centre, Merritt Island, USA
Launch vehicle: Saturn IB (SA-208)
Flight type: Ferry to Earth-orbital space station
Flight time: 84 days 1hr 15min 31sec
Spacecraft weight: About 30,100lb
Crew: Lt-Col Gerald Paul Carr, 41, USMC, commander
Edward George Gibson, 37, science pilot
Lt-Col William Reid Pogue, 43, USAF, pilot

The launch of Skylab 4 was delayed for two reasons, one positive, the other negative. Lift-off was put back to ensure that the astronauts would be in space to observe Kohoutek, being trumpeted as the "Comet of the Century," but also because hairline cracks had been discovered on the fins of the 1964-built Saturn IB. The fins were duly replaced, and at 9.01 am on November 16 Skylab 4 lifted off spectacularly into the clearest sky that many observers could recall. Comet Kohoutek was not so spectacular, although the astronauts did see and photograph it.

All the astronauts on Skylab 4 were rookies. The commander, Gerald Carr, was born in Denver, Colorado, on August 22, 1932, and accumulated one degree in mechanical engineering and two in aeronautical engineering. Carr joined the Navy in 1949 and entered flight training with the US Marines in 1954. At the time of his selection as an astronaut in 1966 he was testing Marine tactical systems. He served on the support crews for Apollos 8 and 12. Among his other distinctions was that of being the father of two sets of twins among his six children.

Skylab 4 is launched from Pad 39B. Built to launch the much larger Saturn V, the pad was modified for the Saturn IB by the addition of a pedestal (Nasa)

Left to right *Skylab 4 crew Carr, Gibson and Pogue (Nasa)*

Science astronaut Edward Gibson pictured during the final Skylab EVA (Nasa via Astro Information Service)

Science pilot was a doctor of engineering physics, Edward Gibson, who had joined Nasa as a 1965 scientist-astronaut. He came from Buffalo, New York, where he was born on November 8, 1936. Gibson served on the support crew for Apollo 12.

Ex-Thunderbirds display team member Bill Pogue was the Skylab 4 pilot. Born in Okemah, Oklahoma, on January 23, 1930, he gained science and maths degrees and flew in combat during the Korean War. He was a Thunderbird from 1955 to 1957 and completed a two-year tour as a test pilot with the British Ministry of Aviation after graduating from the Empire Test Pilots School at Farnborough. Pogue served on the support crews for Apollos 7, 11 and 14 after becoming an astronaut in 1966.

The astronauts docked successfully at the second attempt but then that unpredictable nuisance, motion sickness, struck Bill Pogue before they had even left the command module. The crew did not report this to the ground, and even discussed how to dispose of the evidence, Pogue's sick bag. They did not however realise that their conversation was being recorded on data-storage tapes in the command module. These were automatically played to the ground, where mission controllers heard every detail of the debate.

"Well, Bill, I think we had better tell the truth tonight," said Carr. Then, changing his mind, he suggested that the crew dispose of the evidence in the garbage storage system. Gibson agreed: "They're not going to be able to keep track of that. Let's do that because they seem to make a big decision between whether you throw up or not." Although in the end they decided to keep the sick bag, astronaut chief Alan Shepard reprimanded Carr: "You made a serious error of judgement." A chastened Carr replied: "Okay, Al, I agree, it was a dumb decision."

After this unhappy start morale hit rock-botton. The astronauts complained of having too much to do, and they made mistakes. Then Carr cleared the air in a heart-to-heart session with ground control, and thereafter things went very well. The crew replenished coolant supplies, repaired an antenna, and corrected faults on the ATM and the gyroscope. Record spacewalks were carried out, Christmas was spent in space and Comet Kohoutek was seen during a 7hr 2min EVA by Carr and Pogue. By the end of the mission Carr's total EVA time was 15hr 48min, Gibson's 15hr 17min and Pogue's 13hr 31min. The crew conducted 56 experiments and 26 science demonstrations, achieved 15 subsystems test objectives, carried out 13 student experiments and studied the Sun for 338hr. Total experimental time was 1,563 hr.

The crew splashed down 175 miles south-west of San Diego on February 4, 1974, after a record 84 days in space. Carr had to make a manually controlled re-entry when he discovered that the command module was incorrectly aligned. Carr, Gibson and Pogue had flown 1,260 orbits and covered 34.5 million miles, and were found to be in excellent shape when examined aboard USS *New Orleans* after the recovery. Their 11.17 am splashdown had been the first not to be covered live on television.

Costing a total of $2,460 million, Skylab generated a massive 513 man-days in space, and among the important results from the 270 experiments carried out were no fewer than 182,000 photographs of the Sun. Gibson said that Skylab had increased man's knowledge of the Sun by 100 per cent. But the end of Skylab also marked the beginning of a frustrating hiatus in the US manned spaceflight programme. Apart from the nine-day Apollo-Soyuz Test Project in 1975, no American would go into space again until 1981. Skylab briefly entered the limelight again during July 1979, when it re-entered the Earth's atmosphere, showering the Australian outback with debris.

Soyuz 13   December 18, 1973   Flight 63

Name: Soyuz 13

Sequence: 63rd astro-flight, 50th spaceflight, 48th Earth orbit

Launch date: December 18, 1973

Launch site: Tyuratam, USSR

Launch vehicle: A2 (SL-4)

Flight type: Earth orbit

Flight time: 7 days 20hr 55min 35sec

Spacecraft weight: About 14,450lb

Crew: Maj Pyotr Ilyich Klimuk, 31, Soviet Air Force, commander
Valentin Vitalyevich Lebedev, 31, flight engineer

A space duo with a difference took to the skies at 2.55 pm Moscow time on December 18, 1973. They were the Soyuz 13 crew, Pyotr Klimuk and Valentin Lebedev, both aged 31. Klimuk was the youngest ever cosmonaut when selected at the age of 23 in 1965, and Lebedev had been in the cosmonaut team for only a year.

Soyuz 13 was equipped to the original, pre-Salyut standard, carrying solar panels and enough expendables for longer, independent flights. The mission was designed to try out some of the scientific equipment which was to have been used on the abortive Salyut 2 and 2B flights. In addition, the US-Soviet agreement to fly an Apollo-Soyuz joint mission meant that there was still a need for a solar-winged Soyuz, and the flight served to bolster America's confidence in the integrity of the Soviet spacecraft.

Klimuk and Lebedev, flying on the 50th space mission (excluding the X-15s), also took part in the first Soviet-US simultaneous flight: Carr, Gibson and Pogue in Skylab 4 were on their 32nd day in space when Soyuz 13 was launched. Klimuk was born on July 10, 1942, in Komarovka near the Byelorussian town of Brest. He attended Air Force College and joined the cosmonauts in 1965. Lebedev was born in Moscow on April 14, 1942, and studied at the Moscow Aviation Institute between 1960 and 1966. He then worked in a design bureau developing and testing spacecraft systems before joining the cosmonauts in 1972.

The Soyuz 13 orbital module was fitted with the Orion 2 celestial telescope and the Oasis 2 protein manufacturer in place of the docking probe and ancillary equipment. It was Lebedev's job to operate these systems. Orion 2, mounted on the outside of the spacecraft, took the first ultra-violet spectrograms of planetary nebula. Several dozen "hot" stars were found in a celestial breeding ground near the star Capella, 47 light-years away. The telescope could perform precise observations in the ultra-violet spectrum of stellar objects down to the 13th magnitude, whereas Skylab could only look at stars down to 7.5 magnitude. The closed-loop Oasis bred two types of bacteria which were used to make proteins in a test of the practicability of food production in space colonies. Some Earth-resources and bio-medical work was also carried out.

Another indication of the easing of the Soviet policy of near-total secrecy about current space activities came with an announcement of the half-way point of the mission. The Russians still had a lot to learn about news management,

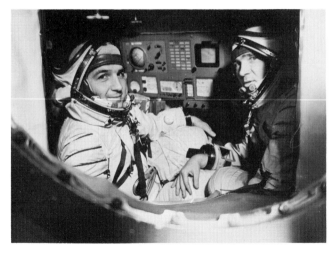

*The 50th officially recognised manned spaceflight was made by Pyotr Klimuk **left** and Valentin Lebedev in Soyuz 13 (Novosti)*

however: the remarkable Geoffrey Perry and his team of schoolboys at Kettering Grammar School, England, announced the end of the flight a good hour before the official Soviet statement. The eight-day mission of Soyuz 13 ended in a snowstorm at 11.50 am on Christmas Eve when the re-entry module touched down 124 miles south-west of Karaganda after 128 orbits of the Earth. Both cosmonauts received the customary honours of Hero of the Soviet Union, the Gold Star and the Order of Lenin.

Soyuz 14   July 3, 1974   Flight 64

Name: Soyuz 14
Sequence: 64th astro-flight, 51st spaceflight, 49th Earth orbit
Launch date: July 3, 1974
Launch site: Tyuratam, USSR
Launch vehicle: A2 (SL-4)
Flight type: Ferry to Earth-orbital space station
Flight time: 15 days 17hr 30min 28sec
Spacecraft weight: About 14,450lb
Crew: Col Pavel Romanovich Popovich, 43, Soviet Air Force, commander
Lt-Col Yuri Petrovich Artyukhin, 44, Soviet Air Force, flight engineer

On June 25, 1974, a giant Proton booster rose ponderously from Tyuratam, taking with it the 40,000lb Salyut 3, a much improved development of the Soviet space station design. The living and working quarters were better, there was a unique system of three articulated, rotatable solar panels which could turn to the sun without disturbing the orientation of the spacecraft, and, most significantly, the station was equipped with an 33ft-focal-length optical telescope, its lenses and photographic system permanently pointing towards Earth. Rumours that the station was a military spying base seemed to be confirmed when the all-military crew

of Col Pavel Popovich and Lt Col Yuri Artyukhin were launched towards it on July 3, 1974, and when code words were used in space-ground communications.

The launch, at about 9.40 pm, placed the Soyuz 14 ferry vehicle about 1,800 miles behind Salyut 3. Two major orbital manoeuvres by Popovich and Artyukhin after Soyuz 14's tenth orbit reduced the distance to 1,000 miles. A further two burns brought Soyuz 14 to within a mile of Salyut 3. At 300ft Popovich took manual control, reducing the approach speed to 1ft/sec at 120ft. Docking was achieved at 9 pm GMT on July 4, 26hr after the launch. The initial soft docking was followed by a hard docking and a thorough check of the pressure seal between Soyuz and Salyut before Artyukhin entered the space base.

Popovich, twelve years after his flight in Vostok 4, was in high spirits. He even used his old callsign, *Golden Eagle.* Artyukhin had joined the cosmonaut team a year after Popovich's flight in 1963, and this was his first assignment. He was born in Pershutino near Moscow on July 22, 1930, and attended Air Force Technical College and the Zhukovsky Engineering Academy.

Salyut 3 had been given a floor and a ceiling of different colours to help the crew to orientate themselves in weightlessness. A strict regime of eight hours on, eight hours off was established. After eight hours of work the crew spent eight hours on exercise, recreation and housekeeping,

*Salutes before **above** and after **below** the flight by the first space spies, Pavel Popovich and Yuri Artyukhin of Soyuz 14. Popovich is on the right in spacesuit and left in uniform (Novosti)*

followed by eight hours for sleep. They used special elasticated suits harnessed to a small running track in an effort to maintain cardiovascular condition. A massive solar storm between July 5 and 8 which sent solar flares and heavy radiation streaming out from the Sun worried the Russians so much that they announced that the crew might have to come home early.

Much was disclosed about the men and their daily regime, but little was said about their activities, chief of which was the use of a powerful telescope and camera system to evaluate the potential of a manned military laboratory. For this purpose test patterns had been laid out on the ground near Tyuratam. If the unclassified Earth-resources photographs taken by the Skylab crews are anything to go by, Popovich and Artyukhin must have had a very interesting and successful time.

After 15 days on board, and three hours 18min after undocking, the crew of Soyuz 14 landed south-east of Dzhezkazgan in Kazakhstan at 3.21 pm. They had completed 252 orbits in 15 days 17hr 30min.

Name: Soyuz 15
Sequence: 65th astro-flight, 52nd spaceflight, 50th Earth orbit
Launch date: August 26, 1974
Launch site: Tyuratam, USSR
Launch vehicle: A2 (SL-4)
Flight type: Ferry to Earth-orbital space station
Flight time: 2 days 0hr 12min 11sec
Spacecraft weight: About 14,450lb
Crew: Lt-Col Gennadi Vasilyevich Sarafanov, 32, Soviet Air Force, commander
Col Lev Stepanovich Demin, 48, Soviet Air Force, flight engineer

*Soyuz 15's crew comprised grandfather Lev Demin **left** and the 32-year-old commander, Gennadi Sarafanov (Novosti)*

The second military Soyuz/Salyut crew, which in 48-year-old Col Lev Demin included the world's first space grandfather, was launched towards Salyut 3 in routine fashion at 10.58 pm on August 26, 1974. Observers assumed that on the 16th orbit the Soyuz 15 ferry would make its port of call in space. They were wrong: on the 16th orbit Soyuz 15 was some 70 miles ahead of Salyut 3 when it should have been making its final approach from behind. Soyuz 15 approached the space station much too quickly in the final moments, closing at something like 30ft/sec after making apparently uncontrollable and excessive burns when about 160ft away. According to the Russians, Soyuz 15 was a manned test of a future unmanned resupply vehicle that would make a totally remote-controlled docking. But why test an unmanned craft with two cosmonauts aboard? It seems far more likely that Soyuz 15 was just a failed standard ferry vehicle that, despite valiant attempts by the crew to make further approaches, ran

out of time and had to come home in the required two days.

Soyuz 15 made an emergency night landing 30 miles south-west of Tselinograd after 32 orbits in almost exactly two days. Because the mission took place during preparations for the Apollo-Soyuz joint flight the Americans were extremely worried by this demonstration of an apparent loss of expertise in simple orbital mechanics. After all, hadn't the Russians automatically docked two Cosmos satellites seven years previously? Nasa was even more dubious when the Russians said that the mission was intended to test different approach procedures and docking methods, and that even if it had docked Soyuz 15 would have come back home anyway. It was also claimed that another objective was a quick comparison of the effects of weightlessness on the two cosmonauts, one of whom was 16 years older than the other. Finally, the night landing was intended to test "evacuating a spacecraft at night".

Mission commander Gennadi Sarafanov was born on January 1, 1942, in Saratov and graduated from military aviation school before joining the cosmonaut team at the age of 23 in 1965. Lev Demin was born in Moscow on January 11, 1926, and studied with the late Vladimir Komarov. He graduated from the Zhukovsky Engineering Academy in 1956 and joined the cosmonauts in 1963.

More evidence to suggest that Salyut 3 was a military space base emerged when a photographic capsule was ejected from the station on September 23 and returned to Earth.

On Christmas Day 1974 the Russians announced that Salyut 3 had completed 2,950 orbits, responded to 800 control commands, been the base for 400 science and technical experiments, and fired its attitude-control thrusters 50,000 times. Having reached the end of its useful life, Salyut 3 was commanded to re-enter the atmosphere in January 1975 and was destroyed.

Soyuz 16   December 2, 1974   Flight 66

Name: Soyuz 16
Sequence: 66th astro-flight, 53rd spaceflight, 51st Earth orbit
Launch date: December 2, 1974
Launch site: Tyuratam, USSR
Launch vehicle: A2 (SL-4)
Flight type: Earth orbit
Flight time: 5 days 22hr 23min 35sec
Spacecraft weight: 14,730lb
Crew: Col Anatoli Vasilyevich Filipchenko, 46, Soviet Air Force, commander.
Nikolai Nikolayevich Rukavishnikov, 42, flight engineer

On May 24, 1972, the United States and the Soviet Union agreed to fly a joint Apollo-Soyuz mission during which the two craft would dock. A lot of detailed work and planning went into the preparation of the flight, which was scheduled for mid-1975. The two spacecraft were to be modified for the mission, Apollo carrying a docking module and both craft fitted with essentially similar docking adaptors which could be used in either the active or the passive modes. All in all, it was going to be a complex mission, conducted in two languages, monitored by two different tracking and communications networks, and calling for agreement on such things as the spacecraft atmosphere to be used during the docking phase. Mindful of their recent string of mishaps and anxious to demonstrate their technological competence to America and the world, the Russians carried out a number of dummy runs. They flew Cosmos 652 and 672 unmanned, followed by Soyuz 16 as an ASTP dress rehearsal.

The crew of Soyuz 16 were the first back-up team for the actual ASTP mission. The Russians announced the names of four crews in May 1973 at the time the Americans named their prime crew and one reserve team. Anatoli Filipchenko and Nikolai Rukavishnikov took Soyuz 16 into space on December 2, 1974, but in spite of the close co-operation prevailing between the Soviets and the US at the time, the 12.40 pm launch was not announced until later, catching the Americans by surprise. This did not add to American confidence in the Russians, and nor did the fact that Soyuz 16 went into the wrong initial orbit, an error explained away as a

ASTP rehearsal crew inside Soyuz 16, with Rukavishnikov in the foreground and mission commander Filipchenko behind (Novosti)

deliberate move to test the Soyuz guidance and orbital manoeuvring systems more fully.

Two major tests were then performed. Using an "imitation docking ring" attached to the front of the Soyuz orbital module, the cosmonauts simulated a variety of docking modes. They also reduced the pressure of the oxygen/nitrogen atmosphere in their cabin from $14lb/in^2$ to $10lb/in^2$ and raised the oxygen content from 20 to 40 percent to match the compromise atmosphere that would be used when the two ASTP craft were docked. During the 17th and 18th revolutions the crew circularised the orbit at 140 miles and signalled to the US to conduct a mock Apollo launch. Important tracking tests were also conducted.

Towards the end of the sixth day the retro-rockets were fired for 166sec, marking the end of what had ultimately turned out to be a satisfactory curtain-raiser for the joint mission. The flight callsign, *Buran* ("snowstorm"), proved very apt, with Soyuz 16 landing on the snow-covered steppes of the prime recovery zone at 11.04 am on December 8. The crew were wrapped in overcoats and whisked away by helicopter to Star City at Tyuratam. The Americans were told of the safe landing five minutes later. The Soyuz 16 ASTP rehearsal mission lasted five days 22hr 24min and totalled 96 orbits.

The androgynous docking system (Novosti)

Soyuz 17   January 11, 1975   Flight 67

Name: Soyuz 17
Sequence: 67th astro-flight, 54th spaceflight, 52nd Earth orbit
Launch date: January 11, 1975
Launch site: Tyuratam, USSR
Launch vehicle: A2 (SL-4)
Flight type: Ferry to Earth-orbital space station
Flight time: 29 days 13hr 19min 45sec
Spacecraft weight: About 14,450lb
Crew: Lt-Col Alexei Alexandrovich Gubarev, 42, Soviet Air Force, commander
Georgi Mikhailovich Grechko, 42, flight engineer

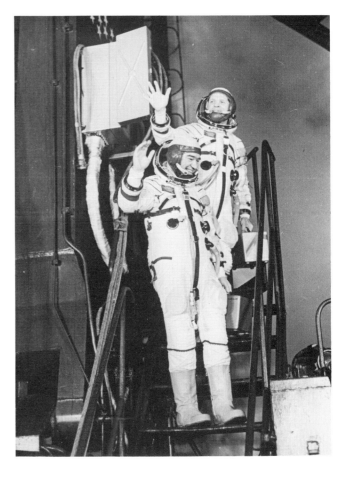

Georgi Grechko **foreground** *and Alexei Gubarev at the Soyuz 17 launch pad* (Novosti)

The Salyut 4 space station was launched on Boxing Day 1974. Weighing nearly 42,000lb, it circled the Earth 220 miles up until on January 12, 1975, it was joined by the 14,000lb Soyuz 17 ferry vehicle, launched at 00.43 am the previous day. When the Soyuz 17 crew, Alexei Gubarev and Georgi Grechko, stepped inside the station they found a notice saying "Wipe your feet," placed there by technicians before the Salyut launch. The cosmonauts connected a hose between the station and the flight cabin of their powered-down Soyuz in order to keep the air inside it fresh. Salyut 4 was essentially similar in configuration to Salyut 3, but unlike its military predecessor was very much a science platform. In fact there was so much scientific and medical equipment on board that it was estimated that the busy flight engineer, Grechko, covered three miles a day floating from instrument to instrument.

An artist's impression of Salyut 3 (Novosti)

his work on the design of the lunar soft-landers, and had served under the late Sergei Korolev, one of the fathers of Soviet space travel. Grechko was born in Leningrad on May 25, 1932, and attended the Leningrad Institute of Mechanics. Joining the cosmonaut team in 1967, he was the back-up research engineer for Soyuz 7.

Gubarev was born in Gvardeitsy, Borsky, on April 29, 1932, and joined the Soviet Army in 1950. He then went to a naval air school and graduated from the Air Force Academy before joining the cosmonauts in 1963. He and Grechko formed the back-up crew for Soyuz 12.

Salyut 4 carried a new self-contained, automatic navigation system called Delta. Based on a radio altimeter, optical instruments and a computer, it calculated the space station's orbital data and engine-firing times. Another interesting item was a water-regeneration system designed to turn into drinkable water evaporated human waste from the space station's atmosphere.

The crew conducted a variety of scientific and biomedical experiments using a bicycle ergometer, endless-belt running track, solar telescope, and the Oasis space "garden". A camera filmed the development of peashoots in weightlessness. The crew saw a supernova in the constellation of Lyra. Each cosmonaut wore a different muscle-loading suit, designated Athlete and Penguin, to evaluate their effectiveness as a means of countering the physiological effects of weightlessness. The solar telescope was a disappointment, the Sun effectively "blinding" its mirrors.

Soyuz 17's crew were quiet, efficient and methodical. The mission was probably the least demonstrative ever flown to that date, and many people were surprised to hear that on their landing Grechko and Gubarev had been in space for nearly 30 days, beating the Soviet record and exceeding the duration of the first Skylab mission. The landing, however, was not without its exciting moments. A wind of 44mph was blowing across the landing site, located 68 miles north-east of Tselinograd, visibility was about 1,600ft and the re-entry module finally emerged from cloud at 800ft. The crew were in excellent shape after their 467 orbits.

Grechko was one of the leading designers in the Soviet space programme. He gained a master's degree in 1967 for

Soyuz 18-1   April 5, 1975   Flight 68

Name: Soyuz 18-1
Sequence: 68th astro-flight, 55th spaceflight
Launch date: April 5, 1975
Launch site: Tyuratam, USSR
Launch vehicle: A2 (SL-4)
Flight type: Sub-orbital abort
Flight time: 21min 27sec
Spacecraft weight: About 14,450lb
Crew: Col Vasili Grigoryevich Lazarev, 47, Soviet Air Force, commander
Oleg Grigoryevich Makarov, 42, flight engineer

Vasili Lazarev and Oleg Makarov, already veterans of the Soyuz 12 mission, boarded their Soyuz 18 ferry vehicle on April 5, 1975, to prepare for a visit to Salyut 4. A record stay of 60 days in the space station was on the agenda. The reliable old A2 booster lifted off from Tyuratam and to begin with all looked well. The strap-on boosters separated, the core stage shut down and the second stage ignited, its exhaust blasting

through the lattice-like framework connecting the core and second stages. The stages were normally separated by the firing of two sets of six pyrotechnic latches, one set at the top and the other at the bottom of the lattice structure. During the launch of Soyuz 18 three of the six latches in the upper section fired prematurely. The malfunctioning latches adjoined one another, so that the upper stage was attached on one half of its circumference and free on the other. An electrical link was broken at the same time, making it impossible for separation to be completed. So, with the upper stage dragging the spent core, the launcher veered from its planned path. The booster's gyroscope detected a deviation of over 10° and the Soyuz was automatically ejected just four seconds after second-stage ignition.

Lazarev and Makarov were powerless to do anything as the automatic abort sequence separated the payload shroud, orbital module and instrument unit and positioned their flight cabin for a low-speed, high-g re-entry from a height of about 90 miles. Only the lifting capability of the Soyuz re-entry

*Soyuz 18-1 crew Lazarev **left** and Makarov (Novosti)*

Seat of the problem that led to the Soyuz 18-1 abort, the lattice framework between the first and second stages is visible in this picture of a later Soyuz T booster (Tass)

module saved the crew from undergoing as much as 18g deceleration. As it was, they were subjected to 14g during the sub-orbital, emergency descent.

The cosmonauts were extremely worried at the time, not necessarily for their safety while airborne but because they thought they were coming down in China. The capsule landed 1,000 miles from Tyuratam in the mountains of Siberia, near Gorno-Altaisk, and 200 miles from China. Lazarev and Makarov climbed out unhurt and lit a fire. They

had to wait a day to be rescued but were found to be in excellent health despite their ordeal, the first ever manned launch abort. An unofficial story states that their re-entry module hit the side of a mountain and rolled downhill.

The failure came just three months before the ASTP mission, and the Americans were naturally concerned about the integrity of the Soviet booster. They were assured that Soyuz 18-1 had been launched by an "old-fashioned" booster and that the ASTP Soyuz would fly on a modernised version.

Soyuz 18   May 24, 1975   Flight 69

Name: Soyuz 18
Sequence: 69th astro-flight, 56th spaceflight, 53rd Earth orbit
Launch date: May 24, 1975
Launch site: Tyuratam, USSR
Launch vehicle: A2 (SL-4)
Flight type: Ferry to Earth-orbital space station
Flight time: 62 days 23hr 20min 8sec
Spacecraft weight: About 14,450lb
Crew: Lt-Col Pyotr Ilyich Klimuk, 32, Soviet Air Force, commander
Vitali Ivanovich Sevastyanov, 39, flight engineer

The Soyuz 18-1 back-up crew, veterans Pyotr Klimuk and Vitali Sevastyanov, safely docked their Soyuz ferry with Salyut 4 during the space station's 2,379th orbit of the Earth

and 24hr after their launch from Tyuratam at 5.58 pm on May 24, 1975. By the fifth day the crew had settled down to work on what was sure to be a long and busy mission.

The schedule on board Salyut was different from that used on the Skylab missions. The Soviets favoured a single-discipline approach under which one area of science would be handled at a time, whereas in Skylab the astronauts had a number of experiments on the go simultaneously. For example, the Soyuz 18 crew carried out multi-spectral photography from June 8 to 11 before moving on to another area of investigation.

The crew took over 2,000 Earth-resources photographs, covering an area of the Soviet Union totalling five million square miles. Sevastyanov gave an indication of the value of these pictures when he claimed after the flight that the

*Pyotr Klimuk **left** and Vitali Sevastyanov prepare to board Soyuz 18B (Novosti)*

"luminescent intensity of atomic oxygen's red line at high altitude". One of the crew's X-ray photographs of the object Cygnus X-1 indicated that it was a black hole. Thirteen days were spent on geophysical work, 13 on astrophysics, six on technical and 10 on medical experiments, two on photography inside the space station, two on atmospheric measurements, and seven on unpacking and packing. The crew also enjoyed a total of 10 days of relaxation.

After 59 days and about 900 orbits in Salyut 4 Klimuk and Sevastyanov returned to Earth. They landed 34 miles northeast of the town of Arkalyk at 5.18 pm Moscow time on July 26. A television recording of their undocking from Salyut 4 had been shown, and their descent was presented live. After the landing the doctors wanted the cosmonauts to be carried from the capsule on stretchers, in order to ease them into their 10-day readaptation to Earth gravity. The cosmonauts refused and Klimuk was first to leave the capsule. A "certain amount of emotional excitation" was experienced by the crew, it was reported afterwards. Klimuk said that it also took some time to readapt psychologically. Days after the flight, he once awoke to see Sevastyanov sleeping with his arms in the air, as if they were floating in zero g.

A "symbolic rendezvous" took place during the Soyuz 18/Salyut 4 mission when the docked Soyuz 19 and Apollo 18 ASTP spacecraft came to within about 200 miles of the space station.

On November 17, 1975, the Russians launched the unmanned Soyuz 20 to dock with Salyut 4 and remain there for 90 days. This was a test of the Progress unmanned tanker that would be used to replenish subsequent space stations. Salyut 4 was not manned again and was commanded to re-enter on February 3, 1977, after completing over 12,000 orbits of the Earth.

whereabouts of even the smallest ore deposits in the Soviet Union was now known.

The crew repaired a cosmic ray instrument and spent a lot of their time operating the solar telescope, with which they took over 600 pictures of the Sun. Other scientific equipment put to use included the KDC short-wave diffractional spectrometer, the Filin and PT4 X-ray telescopes, the Silya isotope spectrometer, the Spectrum cosmic instrument, and the Emission experiment package, which measured the

*Klimuk **right** and Sevastyanov in a Salyut mock-up (Novosti)*

Apollo-Soyuz Test Project

Soyuz 19   July 15, 1975   Flight 70
Apollo 18   July 15, 1975   Flight 71

Name: Soyuz 19
Sequence: 70th astro-flight, 57th spaceflight, 54th Earth orbit
Launch date: July 15, 1975
Launch site: Tyuratam, USSR
Launch vehicle: A2 (SL-4)
Flight type: Earth orbit, rendezvous and docking
Flight time: 5 days 22hr 30min 51sec
Spacecraft weight: 14,729lb
Crew: Col Alexei Archipovich Leonov, 41, Soviet Air Force, commander
Valeri Nikolayevich Kubasov, 40, flight engineer

Name: Apollo 18 (AS-210)
Sequence: 71st astro-flight, 58th spaceflight, 55th Earth orbit
Launch date: July 15, 1975
Launch site: Pad 39B, Kennedy Space Centre, Merritt Island, USA
Launch vehicle: Saturn IB (SA-210)
Flight type: Earth orbit, rendezvous and docking
Flight time: 9 days 1hr 28min 24sec
Spacecraft weight: 32,563lb
Crew: Brig-Gen Thomas Patten Stafford Jr, 44, USAF, commander
Vance DeVoe Brand, 44, command module pilot
Donald Kent Slayton, 51, docking module pilot

On the evening of July, 17, 1975, high over Amsterdam, Holland, 47,000lb of American and Russian spacecraft was orbiting the Earth in a close embrace. Inside were an English-speaking Russian crew of two and three Russian-speaking American astronauts. At the half-way point between the two spacecraft, in the docking module, the two commanders, watched on television by millions of people all over the world,

The international ASTP crew with a mock-up of the docked vehicles:
left to right *Stafford, Brand, Slayton, Leonov and Kubasov (Nasa)*

floated towards each other, their hands outstretched. As Alexei Leonov and Tom Stafford shook hands they said "Glad to see you" in their respective languages. Détente had reach its zenith, with Americans and Russians together in space.

The story of the Apollo-Soyuz Test Project (ASTP) began on May 24, 1972, with the signing of an agreement by President Richard Nixon and Premier Alexei Kosygin. Originally the plan was for an Apollo to dock with a Salyut space station, but this was subsequently changed. A joint flight was attractive to both sides, and with relations between the USA and the Soviet Union at their most cordial ever, the time could not have been more ripe. For the Soviets it was an opportunity to acquire expertise from the Americans and to be seen as on a par with them technically. Equally, it was a chance for America to gain an insight into the Russian programme, and to keep their hand in at manned spaceflight now that Skylab had ended and the Space Shuttle was so far away.

After three years of intensive effort – involving reciprocal visits by teams of scientists and spacemen to Russian and American space centres, the development of a compatible docking adapter, the drawing up of a minutely detailed flight plan, and many rehearsals – the two teams were ready in time for the original target launch date of July 15, 1975.

The Soviet Union got the mission under way with the launch of Soyuz 19 at 3.20 pm Moscow time. The lift-off was shown live across the world, the first time this had happened in the Russian manned programme. On board were veteran cosmonauts Alexei Leonov and Valeri Kubasov, chosen in May 1973. Almost ten thousand miles away, on Pad 39B at the Kennedy Space Centre, Apollo 18 awaited lift-off 7½hr later. On board were Tom Stafford, already with three spaceflights to his credit, and two rookies, one of whom was 51 years old.

The docking module pilot was Deke Slayton, the Mercury astronaut whose cruel fate it was to have been grounded since 1962. When he finally convinced the doctors that he was fit to fly, Mercury, Gemini, Apollo and Skylab had been and gone. ASTP was his very last chance to make a spaceflight after being in the astronaut corps for 16 years. Born in Sparta, Wisconsin, on March 1, 1924, Slayton joined the Air Force, won his wings in April 1943 and flew B-25 bomber missions over Europe and Japan during the Second World War. He gained a science degree in aeronautical engineering at the University of Minnesota and worked for Boeing before rejoining the USAF and attending the Experimental Test Pilot School at Edwards AFB in 1956. He was chosen as a Mercury astronaut on April 2, 1959, and selected in November 1961 for the second American manned orbital spaceflight. Indeed, had a third manned Redstone ballistic mission taken place, Slayton would probably have become the first American in orbit, since the heart condition that resulted in his grounding did not show up until March 1962. Slayton then became chief of the astronauts and was a powerful force in the Nasa hierarchy until March 1972, when he was restored to flight

An impression of the historic handshake (Nasa)

status at the age of 48.

Command module pilot Vance Brand had also had his share of bad luck. He would have orbited the Moon had not budget cuts killed off one of the last lunar Apollo flights. Born in Longmont, Colorado, on May 9, 1931, Brand served in the Marine Corps and as a test pilot for the Lockheed Aircraft Corporation, and gained degrees in business administration and aeronautical engineering. Selected as an astronaut in 1966, he was a support crewman for Apollo 8 and 13, back-up command module pilot for Apollo 15 and back-up commander for Skylab 3 and 4.

"Man, I tell you, this is worth waiting 16 years for," said Slayton as Apollo 18 went into orbit after a successful Saturn IB launch. With the Soyuz acting essentially as a passive target, it was left to the more powerful and versatile Apollo to do the manoeuvring and docking. The first move, though, was for the American spacecraft to turn around and extract the docking module from the S-IVB stage, as was done with the lunar module on Moon-landing missions. The docking

module was an airlock 10ft long and about 3½ft in diameter. At one end was a docking system compatible with that fitted to Soyuz 19; the system at the other end was the standard Apollo arrangement. The Soviet and US docking systems each consisted of an extendable guide ring with three petal-like plates at its circumference, each plate having a capture latch inside it. Once the latches were engaged, the active vehicle retracted its guide ring, pulling the craft together.

During the early stages of the rendezvous the Apollo crew had trouble with the docking probe attaching the command module to the docking module. Curiously, it was the very probe that had caused so much trouble during Apollo 14. The Soyuz crew, meanwhile, were having trouble with their television system. At ASTP ground elapsed time (GET) of T+51hr 49min, after a series of complicated rendezvous manoeuvres, Apollo 18 docked with Soyuz 19 over the Atlantic on the Soviet craft's 36th and the Apollo's 29th orbits. The exchanges in English and Russian between the crews were historic rather than memorable. "Real fine, initiating orientation manoeuvre," Leonov had said in English. "Soyuz is very beautiful," said the Russian-speaking Stafford. "Contact, capture" (Leonov). "We also have capture. We have succeeded, everything is excellent" (Stafford). "Soyuz and Apollo are shaking hands now" (Leonov).

During two days of joint operations, which included live television, talks with US President Gerald Ford and Soviet Premier Leonid Brezhnev, and ceremonial events such as the signing of documents, each astronaut and cosmonaut had a chance to visit the other spacecraft. At one point Brand was inside Soyuz 19, speaking Russian to Soviet television viewers. The crews also conducted a number of limited experiments, including a solar eclipse simulation, and a re-docking exercise.

At GET 102hr 16min, on Apollo 18's 61st orbit and after a number of experiments involving the undocked craft in formation, a final separation burn was performed. As the craft parted Leonov called: "Thank you very much for your big job." The two spacecraft then continued independently, Soyuz 19 landing after 96 orbits at 142hr 32min GET. Watched live by television viewers all over the world, the re-entry module touched down 54 miles north-east of Arkalyk, its descent engines kicking up an enormous dust cloud. Intent on making the most of what was sure to be their last manned flight for a number of years, the Americans kept Apollo 18 in orbit for a while longer. The crew completed a busy scientific programme and affably conducted a live television press conference.

At the end of their 138th orbit the Apollo crew came home. The joint mission had been a success, but disaster struck the American crew as they descended towards recovery ship USS *New Orleans* in the Pacific Ocean 270 miles west of Hawaii. Communications difficulties distracted Brand and he forgot to operate the two Earth landing system switches which would have deployed the parachutes and shut down the thrusters. When the drogue parachute did not come out Brand manually commanded it to deploy, but the swinging of the spacecraft caused the still-armed thrusters to fire to correct the oscillations. Stafford noticed this and shut them down, but by then the thrusters' nitrogen tetroxide propellant was boiling off and entering the cabin via a pressure-relief valve. A significant amount of the highly poisonous gas was

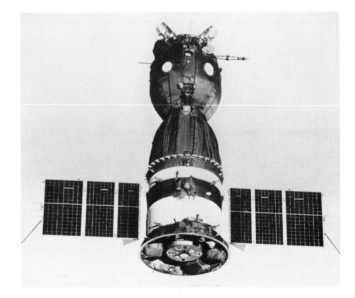

Soyuz 19 in space (Nasa)

drawn into the capsule and the astronauts started to cough and choke. The command module then hit the water "like a ton of bricks," as Stafford described it, and turned upside down. Stafford grabbed oxygen masks from a locker but by the time he reached Brand the command module pilot was unconscious.

None of the astronauts was killed but the fast-acting gas had turned their lungs white and blistered, and during their recuperation a shadow was found on one of Slayton's lungs. Fearing cancer, the surgeons decided to operate but found that it was a benign tumour. Had the shadow shown up before the flight Slayton would probably have been grounded again, which would have been cruel luck indeed.

After the historic joint mission, which cost the USA $218 million, there was much talk of its meaning for the future of mankind and of the possibility of further flights like it. But within months ASTP had been forgotten as the long slide into cold war began all over again.

Soyuz 21   July 6, 1976   Flight 72

Name: Soyuz 21
Sequence: 72nd astro-flight, 59th spaceflight, 56th Earth orbit
Launch date: July 6, 1976
Launch site: Tyuratam, USSR
Launch vehicle: A2 (SL-4)
Flight type: Ferry to Earth-orbital space station
Flight time: 49 days 6hr 23min 32sec
Spacecraft weight: About 14,450lb
Crew: Lt-Col Boris Valentinovich Volynov, 41, Soviet Air Force, commander
Lt-Col Vitali Mikhailovich Zholobov, 39, Soviet Air Force, flight engineer

1976 saw the beginning of a series of Soviet manned flights that were to build up a massive lead over the grounded

Americans, whose Space Shuttle was scheduled for launch in 1978 but in the end didn't make if off the ground until 1981. This period of intense activity was part of the Soviet Government's current five-year plan, which envisaged the "mastery of methods of research from outer space to benefit the national economy". The first moves were the launching of the Salyut 5 space station on June 22 and the subsequent ascent of the Soyuz 21 ferry vehicle at 3.09 pm on July 6, 1976. Aboard were Boris Volynov, a veteran of Soyuz 5, and Vitali Zholobov. A cosmonaut since 1963, Zholobov was born in Zburyevka on June 18, 1937, and after working in the oil and chemical industry joined the Army and then the Air Force. Their activities were to be as much military as scientific in nature, for Salyut 5 was a spy platform as well as a scientific laboratory.

The smoothest Salyut docking to date was achieved on

Soyuz 21's 18th orbit, and the crew set to work. Although much press coverage was given to the science equipment on board, there was little said about the cosmonauts' activities. But the use of code words, the orbit and the nature of the crew all pointed to the fact that theirs was a military mission, again designed to assess the reconnaissance ability of men in space. During their extended stay in orbit the cosmonauts monitored the massive air and sea movements in Siberia during the Operation Sevier military manoeuvres. A tele-printer on board was used to return information to the ground.

The Salyut 5 crew also performed a number of interesting scientific experiments. An aquarium on board allowed them to observe the behaviour and breeding of fish in weightless-ness. The two cosmonauts carried out Earth-resources photography, monitored aerosol and industrial pollution in the upper atmosphere, and conducted time and motion studies. Four other activities were of special interest: crystals were grown in weightlessness; the Potok experiment observed the movement of liquids in weightlessness as a prelude to the development of in-orbit refuelling; the Sphere system was used to melt and harden metals; and the Reaction system was used for soldering.

All appeared to be going smoothly, with the crew apparently set to break the 63-day Soyuz 18B record, when suddenly, 49 days into the mission, they evacuated Salyut 5 and made an emergency return to Earth, landing in darkness 124 miles south-west of Kokchetav on August 25. An acrid odour had been emitted by the environmental control system in the space station. The crew endured if for a while but eventually the atmosphere became unbearable. There were some doubts as to whether Salyut 5 could be used again, and in any event it looked as if the next crew would have no small repair job on their hands.

*Ready to practice emergency splashdown and egress techniques are Soyuz 21 crew Vitali Zholobov **left** and Boris Volynov (Novosti)*

Soyuz 22 September 15, 1976 Flight 73

Name: Soyuz 22
Sequence: 73rd astro-flight, 60th spaceflight, 57th Earth orbit
Launch date: September 15, 1976
Launch site: Tyuratam, USSR
Launch vehicle: A2 (SL-4)
Flight type: Earth orbit
Flight time: 7 days 21hr 52min 17sec
Spacecraft weight: About 14,450lb
Crew: Col Valeri Fyodorovich Bykovsky, 42, Air Force, commander
Vladimir Viktorovich Aksyonov, 41, flight engineer

When ASTP was over the back-up Soyuz for the mission was modified for a special independent flight during which foreign equipment would for the first time be carried on a Soviet manned spacecraft. The ASTP androgynous docking equipment was replaced by a photography canister housing an East German Karl Zeiss MKF-6 multispectral camera. This was the principal tool for the Raduga experiment, designed to "check and improve scientific and technical methods and means of studying geological features of the Earth's surface,

in the interests of the national economies of the Soviet Union and East Germany".

Soyuz 22 entered orbit following a launch at 12.48 pm on September 15, 1976, just 21 days after the return of Soyuz 21 and one day after the Soviets had announced that they were soon to start the Intercosmos series of missions, with flight engineers from socialist countries flying to Salyut space stations. (Nasa had just announced that Western Europeans would eventually fly on the Space Shuttle.) On board Soyuz 22 were Valeri Bykovsky, who had last been in space in 1963, and flight engineer Vladimir Aksyonov. Born on February 1, 1935, in Biglitsy, Aksyonov had been an Air Force officer before becoming an aerospace designer and joining the cosmonauts in 1973.

The 450lb MKF-6 was designed to take multi-spectral photographs of the Earth. At each exposure the system took six high-quality simultaneous pictures of the Earth's surface, using six different cameras fitted with a forward-motion compensation drive. The pictures, taken in the visible and infra-red bands, provided a stereo-imaging capability that was of great use to agricultural experts, cartographers, mineralogists, hydrologists – and intelligence analysts. A

102-mile-wide strip of the surface came under the scrutiny of the cameras at each exposure, and resolution of objects on the ground was about 90ft. Each film cassette inserted into the complex could provide photographs covering six million square miles.

The orbital inclination of Soyuz 22, 65°, took it regularly over East Germany, the northern USSR and the coast of Norway, where, coincidentally, a Nato exercise was being conducted. Pictures of the Baikal-Amur railway construction area assisted in identifying the best routes. Optimum timber production and processing sites were selected, and tidal zones mapped in detail to assist in the siting and design of tidal power stations. Other Earth-resources activities covered by Raduga included studies of mineral prospects in the continental shelf, harvest projections, estimates of land-reclamation possibilities and atmospheric pollution monitoring. In all, Soyuz 22 took 2,400 pictures and covered 30 specific targets in 24 sessions. Bykovsky and Aksyonov came back to Earth after 127 orbits in almost eight days, landing 93 miles north-west of Tselinograd.

Post-landing interview for Soyuz 22's Vladimir Aksyonov **left** *and Valeri Bykovsky* (Novosti)

Soyuz 23   October 14, 1976   Flight 74

Name: Soyuz 23
Sequence: 74th astro-flight, 61st spaceflight, 58th Earth orbit
Launch date: October 14, 1976
Launch site: Tyuratam, USSR
Launch vehicle: A2 (SL-4)
Flight type: Ferry to Earth-orbital space station
Flight time: 2 days 0hr 6min 35sec
Spacecraft weight: About 14,450lb
Crew: Lt-Col Vyacheslav Dmitriyevich Zudov, 34, Soviet Air Force, commander
Lt-Col Valeri Ilyich Rozhdestvensky, 37, Soviet Naval Air Force, flight engineer

The Soyuz 15 crew ran into trouble when trying to board Salyut 3, and had to make an emergency return to Earth. Vyacheslav Zudov and Valeri Rozhdestvensky, flying Soyuz 23 to Salyut 5, didn't fare much better. They too made an emergency return, and earned the unwanted distinction of making the first, inadvertent, Soviet splashdown.

Zudov, like his flight engineer, had been a cosmonaut for 11 years. He was born in Bor, near Gorky, on January 8, 1942, graduated from military aviation school and was an experienced parachutist. Rozhdestvensky, born February 13, 1939, in Leningrad, went to naval engineering school and became a diver in the rescue services of the Baltic Fleet. In a prophetic remark, he said before the flight that his diving experience might "come in handy".

Launched at 8.40 pm, the Soyuz ferry vehicle, with its limited life in space of 2½ days, entered orbit without incident but then suffered a failure of the rendezvous approach electronics. The crew were unable to take manual control, leaving them with no choice but to abandon the mission. The happy-go-lucky, cheerful pair fell silent, all systems, including the radio, having been powered down to eke out the ferry craft's limited supplies of consumables until its orbit brought Soyuz 23 over the recovery zone once more. It was the seventh time in 11 attempts that the Russians had failed to dock.

The craft began its emergency landing on the 33rd orbit. The 20-mile-wide Lake Tengiz was located in the recovery zone, 121 miles south-west of Tselinograd, and bad luck took the Soyuz 23 descent module right into it. To make matters worse, it was night-time, there were high winds and the temperature was about –20° C. The lake was practically frozen over, making a rescue by rafts and a ground crew impossible. Several helicopters located the craft and shone searchlights on it, while another towed it ashore. According to the Russians, a certain amount of heroism was involved in what was obviously a dramatic rescue. The mission had lasted just six minutes over two days.

It was left to Zudov and Rozhdestvensky's back-ups in Soyuz 24 to make the next docking with Salyut 5. But that had to wait until the following spring, when the fierce Kazakhstan winter had relaxed its grip on the launch and recovery zones.

Soyuz 23 crew Vyacheslav Zudov **left** *and Valeri Rozhdestvensky* (Novosti)

Name: Soyuz 24
Sequence: 75th astro-flight, 62nd spaceflight, 59th Earth orbit
Launch date: February 7, 1977
Launch site: Tyuratam, USSR
Launch vehicle: A2 (SL-4)
Flight type: Ferry to Earth-orbital space station
Flight time: 17 days 17hr 25min 58sec
Spacecraft weight: About 14,450lb
Crew: Col Viktor Vasilyevich Gorbatko, 42, Soviet Air Force, commander
Lt-Col Yuri Nikolayevich Glazkov, 37, Soviet Air Force, flight engineer

Where the Soyuz 23 crew had failed, their reserves, Viktor Gorbatko and Yuri Glazkov, succeeded. On February 8, 1977, on the 19th orbit of Soyuz 24, the cosmonauts made contact with the Salyut 5 space station at an approach speed of 1ft/sec. After their launch at 7.12 pm Moscow time on February 7 the Soyuz 24 cosmonauts had sat back while four orbital corrections were made under ground control, taking them to within 5,000ft of Salyut 5 at an approach speed of 6.5ft/sec. The space station was illuminated by Soyuz 24's floodlights as, at 262ft, manual control was taken. Glazkov was first inside and found conditions "comfortable," indicating that the atmospheric problems encountered by the Soyuz 21 crew had been overcome.

Gorbatko was a veteran of Soyuz 7 and, as a back-up on Soyuz 5, had undergone intensive spacewalk training. Glazkov was also a trained spacewalker, and it came as a surprise to Western observers when their mission ended without any attempt at EVA. Glazkov became a cosmonaut in 1965 and, as well as specialising in spacewalking procedures, had served as a ground controller on previous Soyuz flights. He was born in Moscow on October 2, 1939, and graduated from the Air Force Higher Engineering College in 1962.

A range of experiments were carried out on the Soyuz 24/Salyut 5 flight. Soldering trials proved a success, but attempts at casting metals were not so rewarding. The cosmonauts conducted crystal growth experiments, observed fungi and fish roe development in zero g, photographed the Sun, carried out Earth-resources photography, and made a study of glacial precipitation based on observations by the Soyuz 21 crew. They also repaired an onboard computer.

As part of the continuing Soviet investigation of the effects of weightlessness on the human body the cosmonauts used the Polinom 2M system to carry out full electrocardiogram monitoring, and in an interesting investigation looked into the "ceiling of sensitivity of the vestibular apparatus to electric irritants under conditions of weightlessness".

Another significant event was the replacement, for the first time ever, of a proportion of the space station's atmosphere. On February 21 Glazkov, watched by television cameras, opened a valve to vent part of the atmosphere, which was immediately replaced by compressed air from on-board containers.

After a comparatively short flight of 17 days, no doubt necessitated by the crew's military work and the need to return close-up reconnaissance photos, the cosmonauts landed in high winds, low cloud and sub-zero temperatures 23 miles north-east of Arkalyk. They had completed 286 revolutions of the Earth. Two days later, what was presumed to be a military photographic capsule was ejected from Salyut and returned to Earth. Salyut 5 survived until August 8, 1977, when it was commanded to re-enter the atmosphere. Its place was taken by the incredible Salyut 6, the setting for a five-year epic in Soviet space history.

*Pictured with Alexei Leonov and an A2 second stage are Soyuz 24's Viktor Gorbatko **centre** and Yuri Glazkov **right** (Novosti)*

Soyuz 24 on the pad (Novosti)

Name: Soyuz 25
Sequence: 76th astro-flight, 63rd spaceflight, 60th Earth orbit
Launch date: October 9, 1977
Launch site: Tyuratam, USSR
Launch vehicle: A2 (SL-4)
Flight type: Ferry to Earth-orbital space station
Flight time: 2 days 0hr 44min 45sec
Spacecraft weight: About 14,450lb
Crew: Lt-Col Vladimir Vasilyevich Kovalyonok, 35, Soviet Air Force, commander
Valeri Viktorovich Ryumin, 38, flight engineer

On September 29, 1977, a massive D-1 booster launched Salyut 6 into orbit, marking the start of celebrations of the 20th anniversary of Sputnik 1 in customary spectacular style. Ten days later, the Sputnik 1 pad at Tyuratam was swarming with techicians preparing the A2 booster that would launch Soyuz 25. Before they entered the spacecraft the two cosmonauts could be heard making final speeches during the course of an unprecedented pre-launch television programme. The flight engineer, Valery Ryumin, hailed the flight as the beginning of the third decade of space. Vladimir Kovalyonok, the commander, said: "We're off to work and in a cheerful mood. . ." The cosmonauts, and Soviet leaders and scientists, were not in such an expansive mood two days later.

Kovalyonok became a cosmonaut in 1967 and was the Soyuz 18B back-up commander. He was born on March 3, 1942, in Byeloye, near Minsk, and attended Air Force School before flying transport aircraft and becoming a paratroop instructor. Ryumin was an electronics engineer who helped design spacecraft and Salyut 6 itself. He was born in Komsomolsk-on-Amur, Siberia, on August 16, 1939, and

joined the cosmonauts in 1973, acting as a capcom during ASTP.

Soviet television viewers were able to see the cosmonauts in the Soyuz cabin during the 5.40 am launch, which was less than spectacular because the rocket disappeared into cloud within 15sec. Once in orbit, Soyuz 25 was 27min and 9,000 miles behind its target, Salyut 6. Western observers expected the cosmonauts to spend a marathon spell aboard the space station, possibly breaking America's world record, and to make a spacewalk together to celebrate the anniversary of the Revolution on November 7.

Trajectory corrections were made until on Soyuz 25's 18th orbit its path matched that of Salyut 6 and the docking manoeuvre began. The Kettering Grammar School team, led by Geoffrey Perry, monitored the first stages of the standard approach and then hit a wall of radio silence. Many hours later the Russians announced that Soyuz 25 was returning to Earth. "From a distance of 396ft," they said, "the docking manoeuvre was carried out, but due to deviations the flight was called off."

There are indications that Soyuz 25 completed a soft docking. According to Western observers, Soyuz 25 was still with Salyut 6 on its 23rd orbit and made two further, unsuccessful, attempts at a hard docking. The ferry vehicle then had to be powered down before making the standard emergency return after 32 orbits. The re-entry module landed 115 miles from Tselinograd on October 11. It had been the eighth Russian space station-related mission to fail in 13 attempts.

The Soviets clearly feared that the Soyuz 25 failure meant that there was a fault in the docking system on Salyut 6. But even if it did, they still had another card to play, as the next Soyuz mission proved.

A commander and his flight engineer: Soyuz 25's Vladimir Kovalyonok **left** *and Valeri Ryumin* (Tass)

Name: Soyuz 26
Sequence: 77th astro-flight, 64th spaceflight, 61st Earth orbit
Launch date: December 10, 1977
Launch site: Tyuratam, USSR
Launch vehicle: A2 (SL-4)
Flight type: Ferry to Earth-orbital space station
Flight time: 96 days 10hr 0min 7sec (landed in Soyuz 27)
Spacecraft weight: About 14,450lb
Crew: Lt-Col Yuri Viktorovich Romanenko, 33, Soviet Air Force, commander
Georgi Mikhailovich Grechko, 45, flight engineer

A day after the 4.19 am launching of Soyuz 26, cosmonaut Yuri Romanenko took manual control of his space ferry and docked with Salyut 6 "from only a few metres," indicating a change in procedures. More surprising, however, was the fact that Romanenko and Georgi Grechko had docked at a second port at the rear of Salyut 6. If there was indeed a fault in the primary docking port, as was indicated but not proved by the Soyuz 25 failure, then the existence of a second port, hitherto unsuspected by Western observers, was the saving of the space station.

The two cosmonauts, who were clearly scheduled for a marathon stay in orbit, were 45-year-old Soyuz 17 veteran Georgi Grechko and the 33-year-old commander, Yuri Romanenko, who was one of the four cosmonauts to have been named before they flew, having been nominated as the second reserve for ASTP-Soyuz 19. Romanenko was born on August 1, 1944, in the Orenburg settlement of Koltubanovsky and went to air force college, became a flight instructor and joined the cosmonauts in 1970.

*Yuri Romanenko **left** and Georgi Grechko practice inside a Salyut ground trainer (Novosti)*

After docking on orbit 21 the Soyuz 26 crew spent some hours checking systems before Grechko went inside the Salyut. As Romanenko followed, Grechko turned the television camera towards him and viewers heard him say: "Come on, show yourself." Their Salyut 6 home was much improved, with a temperature regulation and water reclamation system, computer, teleprinter, navigation system, external television cameras, airlock, EVA handrails, sun sensor, waste ejection airlock, dust filter, running track, ion sensor, the MKF-6 multi-spectral camera, and a shower. It was 70ft long and had three rotating solar panels which could be extended an additional nine feet when required.

The cosmonauts, working to a new schedule linked to Moscow time and at a reduced rate, conducted many experiments and were featured many times on live television, during which they gave enthusiastic running commentaries. On December 19 Grechko made the first Soviet spacewalk for nine years, going out to inspect the suspect prime docking port. He found it in full working order and ready to receive a visit from another Soyuz. Grechko presented an excellent television show, using a hand-held colour camera, and he and Romanenko, who was exposed to the vacuum of space while in the airlock, wore new semi-rigid spacesuits incorporating portable life-support systems. Grechko's EVA lasted about 20min, although the time from depressurisation to repressurisation was 88min. The cosmonaut worked in sunlight at all times, and was exposed to a maximum temperature of 316°F. In a duplication of American practice, training for the EVA had been carried out in transport aircraft flying parabolic curves, and in a neutral buoyancy tank. During the EVA Romanenko, keen to get as far outside as he could, lost his grip and floated out of the airlock. Only Grechko's instinctive grab saved Romanenko from a lingering death in space.

After the cosmonauts had celebrated Christmas – there was a real fir tree on board – and the New Year, the next big event on their calendar was a visit from Soyuz 27. The first space visitors arrived on January 11, 1978, docked at the prime port and stayed for eight days. They returned to Earth in Soyuz 26, leaving the fresh Soyuz for use by the long-staying crew.

The most technically significant part of the flight came on January 20, 1978, when Progress 1 was launched. This unmanned ferry vehicle, weighing 15,444lb and carrying propellants, compressed air, food, water, films, letters, parcels and scientific equipment, docked at the rear port of Salyut 6 and on February 2 acted as tanker for the first refuelling in space. Still very much operational, Progress is essentially a stripped-down Soyuz, lacking descent and landing systems, which can fly independently for eight days or docked to a Salyut for a month. At the end of its stay, on February 6, Progress 1 was loaded with waste and rubbish before backing away to a distance of ten miles, making a test automatic approach, and deliberately re-entering to destruction over the Indian Ocean.

One of the pieces of equipment brought up to Salyut 6 by Progress 1 was a solar electric furnace capable of temperatures of up to 1,000°C and used to study the diffusion processes in molten metals and the interaction of solid and liquid metals in weightlessness. Another interesting experiment conducted by Romanenko and Grechko was Medusa, in which amino acids and other biological building blocks were exposed to radiation.

Soyuz 28 was launched on March 2, 1978, and visited Salyut 6 as Romanenko and Grechko were breaking the American-held space duration record of 84 days. As the visitors left after an eight-day mission the Russians announced that the record-breaking space hosts would be coming home soon. This they did after a flight of 96 days 10hr, making Grechko the individual duration record-holder, with a total of 125 days 23hr 20min to his credit. Soyuz 27

landed 165 miles west of Tselinograd. "We are very glad to be back on our planet we love so much, our Mother Earth", said Romanenko.

The two cosmonauts took a long time to readapt to 1g. As a Russian scientist said some days after their landing: "They are both still up there." But readapt they did, demonstrating that the limit to the time men could spend in space had not yet been reached. In a more than usually fulsome tribute, the Russians said that this achievement was due in a large part to the cosmonauts' "sense of humour, their enthusiasm for work and their mutual trust and understanding".

Soyuz 27   January 10, 1978   Flight 78

Name: Soyuz 27
Sequence: 78th astro-flight, 65th spaceflight, 62nd Earth orbit
Launch date: January 10, 1978
Launch site: Tyuratam, USSR
Launch vehicle: A2 (SL-4)
Flight type: Ferry to Earth-orbital space station
Flight time: 5 days 22hr 58min 58sec (landed in Soyuz 26)
Spacecraft weight: About 14,450lb
Crew: Lt-Col Vladimir Alexandrovich Dzhanibekov, 35, Soviet Air Force, commander
Oleg Grigoryevich Makarov 45, flight engineer

At 3.26 pm on January 10, 1978, the Salyut 6/Soyuz 26 complex was 17min past Tyuratam when the engines of another A2 launcher ignited. The A2's payload was the first manned resupply craft, Soyuz 27, carrying cosmonauts Vladimir Dzhanibekov and Oleg Makarov. Dzhanibekov, the commander, was born on May 13, 1942, in Iskander and studied physics at Leningrad University before joining a flying club and entering pilot training in a military flying school. He was also a radio and electronics expert. Selected as a cosmonaut in 1970, he served as an ASTP back-up pilot. The frail-looking Makarov was making his second flight into orbit and his third spaceflight in all.

Salyut 6 crew Romanenko and Grechko retreated into Soyuz 26 for safety reasons before the arrival of Dzhanibekov and Makarov on their 17th orbit. The station was manoeuvred for the docking, which could well have been totally automatic because there were none of the customary references to the commander taking manual control. Though the Russians had feared that the addition of a third component might overstress the Salyut-Soyuz combination, their worries proved groundless. Within three hours of the docking Makarov and then Dzhanibekov were being dragged inside by the enthusiastic Romanenko and Grechko and greeted with bear hugs and kisses. The cosmonauts then toasted their success in tubes of cherry juice during a 15min live television show.

Three space firsts had been achieved: the first four-man space station, the first dual docking and the first resupply. The Soyuz 27 crew handed over copies of the newspaper *Pravda*, letters, books and research equipment. It was announced that their stay in Salyut 6 would only last five days and that the cosmonauts would return in Soyuz 26, leaving the fresh Soyuz 27 for the marathon crew and freeing the rear docking port for the unmanned Progress 1 tanker.

*Soyuz 27's Vladimir Dzhanibekov **left** and Oleg Makarov. The latter was the tenth man to carry out three space missions (Novosti)*

Experiments conducted primarily by the Soyuz 27 crew during their stay included the Soviet-French Cyton biology experiment, designed to study the effects of radiation on cell division; research into the circulatory system to obtain data on blood redistribution and cardiac activity in zero g; and the Resonance experiment, in which a cosmonaut jumped on the "floor" of the space station to check the effects of such loads on the total docked system, which was now 120ft long and weighed about 70,000lb. Dzhanibekov also gave the space station's electrical system a thorough check.

Preparing for their return to Earth in Soyuz 26, the Soyuz 27 crew swapped cabin couches and packed the descent module with film, packages and other material. Dzhanibekov and Makarov landed 192 miles west of Tselinograd at 11.25 GMT on January 16 and were greeted with bread and salt, the traditional Russian welcome for travellers. The way was now open for the Intercosmos team of cosmonauts to visit Salyut 6, and the first was not long in coming.

Soyuz 28   March 2, 1978   Flight 79

Name: Soyuz 28
Sequence: 79th astro-flight, 66th spaceflight, 63rd Earth orbit
Launch date: March 2, 1978
Launch site: Tyuratam, USSR
Launch vehicle: A2 (SL-4)
Flight type: Ferry to Earth-orbital space station
Flight time: 7 days 22hr 16min
Spacecraft weight: About 14,450lb
Crew: Col Alexei Alexandrovich Gubarev, 45, Soviet Air Force, commander
Capt Vladimir Remek, 29, Czech Army Air Force, cosmonaut researcher

The political nature of the Soyuz 28 flight overshadowed the fact that the 95th man to fly above 50 miles, the 87th to fly into space and 87th in Earth orbit was not a Russian or an American. He was a Czechoslovakian army pilot called Vladimir Remek.

His historic flight to and stay in Salyut 6 was accompanied by enough propaganda for 50 missions. "A culmination of Czechoslovakia's participation in the Intercosmos programme . . . an expression of confidence and recognition of Czechoslovakia's science and people by the Soviet Union," said Yaroslav Kozesnik, President of the Czechoslovak Academy of Science. "The bloody fighting against Hitlerite fascism ended with the liberation of fraternal Czechoslovakia, and today a Czech citizen is the first to work with the Soviet Union to use cosmic space and obtain scientific research findings for peaceful purposes," said cosmonaut chief Georgi Beregovoi. And so it went on, with nobody mentioning the fact that the Soviet Union had invaded Czechoslovakia ten years before.

Remek was one of the two Czech pilots who, along with two Poles and two East Germans, were selected for the initial Intercosmos flights. All six were regarded as having an equal chance to make the first flight, and Remek was the lucky man. His Czech back-up was Oldrzich Pelczak, who was teamed with Soviet cosmonaut Nikolai Rukavishnikov. The decision was made just a month before the 6.28 pm launch on March 2, 1978, which was watched by many Czech officials at Tyuratam. Commander of Soyuz 28 was Soyuz 17 veteran Alexei Gubarev.

Born on September 26, 1948, in Ceske-Budejovice, Remek left school in 1966 and entered higher air school before being sent to the Soviet Union in 1972 to study at the Gagarin Air Academy. He returned to Czechoslovakia to serve as a pilot in the army and then went to Russia in 1976 as an Intercosmos pilot candidate. He and the other five candidates entered training for Soyuz 28 in August 1977. The training of the cosmonaut researchers was not as intense as that of the Soviet commanders, and each was given a number of scientific tasks based in equipment provided by his home country.

A television camera on Soyuz 28 showed viewers the docking with Salyut 6, and soon Remek, Gubarev, Romanenko and Grechko were celebrating with a degree of enthusiasm that did not abate for many hours. In fact the cosmonauts were told to "sober down" even though they had

Soyuz 28 crewmen Vladimir Remek **upper** *and commander Alexei Gubarev (Novosti)*

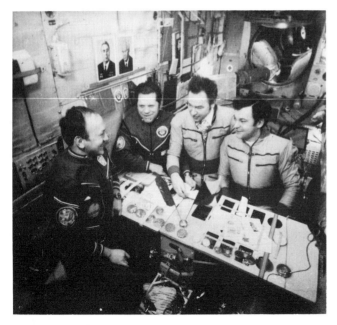

Left to right *Remek, Gubarev, Grechko and Romanenko dine together aboard the orbiting Salyut 6 (Novosti)*

had nothing more powerful than cherry juice to wash down the traditional bread and salt.

The Czech experiments, conducted mainly by Remek, included the Morava investigation, which used the Splav-01 furnace to study the growth of super-pure crystals in zero g and the possibility of obtaining complex semi-conductor and optical materials. Extinctica observed the change in the brightness of stars when viewed through the Earth's atmosphere, Chlorella looked at the growth of algae cultures in a nutrient medium, and an oxymeter was used to study

Soyuz 28 on the pad (Novosti)

oxygen concentrations in human tissue under zero g.

After six days aboard Salyut 6 the international crew of Soyuz 28 landed in a snow-covered field 192 miles west of Tselinograd. Five helicopters and three jets converged on the capsule, and the cosmonauts received an enthusiastic welcome from watching farm workers.

Soyuz 29   June 15, 1978   Flight 80

Name: Soyuz 29
Sequence: 80th astro-flight, 67th spaceflight, 64th Earth orbit
Launch date: June 15, 1978
Launch site: Tyuratam, USSR
Launch vehicle: A2 (SL-4)
Flight type: Ferry to Earth-orbital space station
Flight time: 139 days 14hr 47min 32sec (landed in Soyuz 31)
Spacecraft weight: About 14,450lb
Crew: Col Vladimir Vasilyevich Kovalyonok, 36, Soviet Air Force, commander
Alexander Sergeyevich Ivanchenkov, 37, flight engineer

On June 15, 1978, a new phase in the Salyut 6 saga began with the launching of a new long-stay crew. In the course of their record-breaking 4½-month mission they would play host to two Intercosmos crews and receive three unmanned Progress tankers. The new Salyut 6 lodgers were Soyuz 29 crew Vladimir Kovalyonok, commander of the failed Soyuz 25 and back-up commander for Soyuz 26 and 27, and flight

engineer Alexander Ivanchenkov, who was one of the Soyuz 16 back-ups. Ivanchenkov was born in Ivanteyevka on September 28, 1940, and went to the Moscow Aviation Institute, from which he graduated in 1964. He then worked in a space design bureau and, along with other designers, joined the cosmonauts in 1970.

Soyuz 29 rose from Tyuratam at 11.17 pm and docked at the forward port of Salyut 6 on June 17. The cosmonauts entered the space station and found a welcoming letter from Romanenko and Grechko. They soon got to work, taking the station out of mothballs, replacing a defective ventilator and overhauling an airlock before starting a three-day experiment with the Splav-01 furnace.

Before the flight the crew had received training in ground observation for Earth-resources and reconnaissance purposes, flying in a Tu-134 airliner to about 30,000ft for this work. Putting their training to full use, the cosmonauts took over 18,000 photographs covering much of the Earth's surface. They used the MKF-6 multi-spectral camera and a new topographical camera, the KT-140. An area near Rostov was specially laid out as a test target with grains, grasses and other vegetation.

The crew carried out the usual multitude of experiments, including glass fusing in the Kristall system to obtain data on semi-conductor production techniques. On July 29 Ivanchenkov performed a 2hr 5min spacewalk during which he collected experimental materials samples from the exterior of Salyut 6. They included rubber, polymers and biopolymers, and Soviet scientists were keen to see the effects of space exposure on such materials. Towards the end of the EVA, in which Kovalyonok participated to a lesser extent, ground control said: "If you have nothing more to do then you can get back in." Kovalyonok replied: "We would just like to take our time. This is the first time in 45 days that we have got out into the street to have a walk." Not surprisingly, Kovalyonok and Ivanchenkov proved to be enthusiastic commentators during the spacewalk, which was covered on television.

The routine life on board Salyut 6 included regular meals whose little rituals seemed to be of great importance to the crew. Breakfast might comprise tinned ham, white bread, cottage cheese, blackcurrant jam, cake, coffee and vitamin

*Soyuz 29 crew Vladimir Kovalyonok **upper** and flight engineer Alexander Ivanchenkov (Novosti)*

A cosmonaut practices EVA from a Soyuz mock-up in a neutral-buoyancy tank (Novosti)

pills, followed by a lunch of vegetable soup, cheese, biscuits, tinned chicken, plums, nuts and vitamins, and a dinner of tinned steak, black bread, cocoa and fruit juice.

The cosmonauts received many visits during their stay. First aboard, on June 29, were the Soyuz 30 Intercosmos crew. Then, on July 9, came the Progress 2 unmanned tanker, which remained docked for 25 days and transferred 1,322lb of fuel and oxidiser automatically while the crew worked at other tasks. Progress 3 arrived on August 10 - bringing 617lb of food, 41gal of water, 992lb of oxygen and new processing material - and stayed for 12 days. An East German research cosmonaut and his Russian commander came next on Soyuz 31. Then Progress 4 arrived on October 4, carrying food, fuel and a home comfort for the cosmonauts - fur boots. It returned to Earth on October 26. In a space first, Kovalyonok and Ivanchenkov had moved Soyuz 31 from the rear of Salyut 6 to the front to make room for Progress 4.

On August 8 Salyut 6's total occupation time exceeded the American Skylab's 171 days, and on September 21 the cosmonauts broke 100 days on their own account. Ivanchenkov celebrated his 38th birthday on board on September 28, and on October 9 the Russians announced that the marathon flight would end once the crew had unpacked Progress 4, set up Salyut 6 for another record stay and used the tanker to manoeuvre the station into a convenient orbit for their re-entry in Soyuz 31.

The end of the flight came on November 2 at 2.05 pm Moscow time with a landing 112 miles south-east of Dzhezkazgan. Kovalyonok and Ivanchenkov were surprisingly fit after their record-breaking mission, although they displayed the usual readaptation symptoms during their stay in the rehabilitation centre at Star City.

Soyuz 30   June 27, 1978   Flight 81

Name: Soyuz 30
Sequence: 81st astro-flight, 68th spaceflight, 65th Earth orbit
Launch date: June 27, 1978
Launch site: Tyuratam, USSR
Launch vehicle: A2 (SL-4)
Flight type: Ferry to Earth-orbital space station
Flight time: 7 days 22hr 2min 59sec
Spacecraft weight: About 14,450lb
Crew: Col Pyotr Ilyich Klimuk, 35, Soviet Air Force, commander
Lt-Col Miroslaw Hermaszewski, 36, Polish Air Force, cosmonaut researcher

Polish cosmonaut researcher Miroslaw Hermaszewski displayed a refreshing frankness when he described his return to Earth after his flight to Salyut 6 in Soyuz 30. "The landing was very dynamic, beautiful and fascinating," he said. "You sit in the landing capsule and you see nothing but flames all around you. But you know nothing is going to happen to you, and that you must take it all in, watch everything that happens, to be sure to remember it later. It's beautiful and frightening at the same time, and there is a little doubt and uncertainty. You trust the machines, you trust the technology, but in your heart of hearts you may still have that nagging doubt."

Hermaszewski was born in Lipniki, south-west Poland, on September 15, 1941, and flew gliders at the age of 17. He enrolled in an officers' flying school in 1961, became a fighter pilot in the Polish Air Force in 1964, and by 1976 was commander of a fighter regiment. He was one of two pilots - the other was Col Zenon Jankowski - to be sent to the USSR in December 1976 for the Intercosmos programme. The fact that Jankowski was teamed with Valeri Kubasov was a further indication that the Soviets were turning some of their civilian flight engineers into commanders, Rukavishnikov having been the Soyuz 28 back-up commander.

The Polish cosmonaut's partner was a seasoned veteran, Pyotr Klimuk of Soyuz 13 and 18. They were launched at 6.27 pm on June 27, 1978, and docked at the rear port of Salyut 6 at 8.08 pm on June 29. Materials-processing experiments were high on the list of priorities during their brief stay in Salyut. The Polish Serena experiment called for the fabrication of a cadmium-tellurium-mercury semi-conductor material, the most sensitive known detector of infra-red radiation and worth £4,000 per gram. The rest of the agenda comprised Relaks, a study of the most favourable conditions for rest in weightlessness; the Ziema programme of Earth-resources photography of Poland, partially thwarted by bad weather; Kardiolider, a study of the cardiovascular system; the Zdrowie health experiments; and perhaps the most interesting of all, the Smak investigation of why some foods that are delicious on Earth taste like sawdust in space.

Klimuk and Hermaszewski stayed with Kovalyonok and Ivanchenkov for six days 17hr before departing in Soyuz 30. They landed on July 5 in a maize field in Rostov, 186 miles west of Tselinograd.

Soyuz 30 crew Pyotr Klimuk **upper** *and Miroslaw Hermaszewski* (Novosti)

Soyuz 31   August 26, 1978   Flight 82

Name: Soyuz 31
Sequence: 82nd astro-flight, 69th spaceflight, 66th Earth orbit
Launch date: August 26, 1978
Launch site: Tyuratam, USSR
Launch vehicle: A2 (SL-4)
Flight type: Ferry to Earth-orbital space station
Flight time: 7 days 20hr 49min 4sec (landed in Soyuz 29)
Spacecraft weight: About 14,450lb
Crew: Col Valeri Fyodorovich Bykovsky, 44, Soviet Air Force, commander
Lt-Col Sigmund Jähn, 41, East German Air Force, cosmonaut researcher

Spectacular night launch of Soyuz 31 (Novosti)

The flight plan for the third Intercosmos mission to Salyut 6 was, on the face of it, rather lacking in substance. Experiments called Andro, Berolina, Vremya and Reporter sounded impressive, but it seems unlikely that East German cosmonaut Sigmund Jähn was pushed too hard during the Soyuz 31/29 flight. Jähn and his commander, Vostok 5 and Soyuz 22 veteran Valeri Bykovsky, studied crystal growth and photographic methods, performed a test to distinguish "subtle nuances of sound," and carried out a reaction experiment. The German cosmonaut was also chief photographer, using the MKF-6M Earth-resources camera during a major military exercise.

Born in Rahtenbranz, now East Germany, on February 13, 1937, Jähn graduated from military flight school before entering the Gagarin Air Force Academy in 1966. He became an Intercosmos pilot in 1976. Back-up crew for Soyuz 31 were Viktor Gorbatko, of Soyuz 7 and 24, and Eberhard Kollner.

Soyuz 31 docked with Salyut 6 over Lake Dalkhash. Kovalyonok was waiting eagerly for his new visitors: "We are looking forward to a joyful moment," he said. Jähn entered first and was engulfed in bear hugs. Bykovsky followed, saying: "We were firmly determined to join you and we were convinced that we'd be meeting you here today. Do you remember how we rehearsed everything back on Earth? Well, what am I to say about my feelings at this moment? You know you are real heroes, having stopped up for so long." Jähn then surprised his commander and the "space heroes" by presenting them with souvenir watches.

Soyuz 29 was used for the Intercosmos crew's return. The descent capsule landing 87 miles south-east of Dzhezkazgan on September 3 after a flight of seven days 20hr 49min.

*Valeri Bykovsky **left** and Sigmund Jähn of Soyuz 31* (Novosti)

*Ryumin **left** and Lyakhov are checked up immediately after their 175-day flight (Novosti)*

Name: Soyuz 32
Sequence: 83rd astro-flight, 70th spaceflight, 67th Earth orbit
Launch date: February 25, 1979
Launch site: Tyuratam, USSR
Launch vehicle: A2 (SL-4)
Flight type: Ferry to Earth-orbital space station
Flight time: 175 days 0hr 35min 37sec (landed in Soyuz 34)
Spacecraft weight: About 14,550lb
Crew: Lt-Col Vladimir Afanasevich Lyakhov, 37, Soviet Air Force, commander
Valeri Viktorovich Ryumin, 39, flight engineer

Soyuz 34 descends to Earth carrying the Soyuz 32 record-setters (Novosti)

The crew of Soyuz 32 were looking forward to hosting three Intercosmos pairs in Salyut 6 when they were launched at 2.54 pm on February 25, 1979. But as fate would have it, Vladimir Lyakhov and Valeri Ryumin were destined to spend 175 long, lonely days together. This was Ryumin's second flight and Lyakhov's first. Born on July 20, 1941, in Antratsit, Lyakhov attended the Kharkov Higher Air Force School and joined the Air Force in 1964. He joined the cosmonauts in 1967, graduated from the Gagarin Air Force Academy in 1975 and subsequently served as Soyuz 29 back-up commander.

Soyuz 32 docked with Salyut 6 at 4.30 pm on February 26. The first visitor to the space station was Progress 5, launched on March 12. The supply craft delivered a videotape recorder and monitor to permit two-way visual communications, and a redesigned Kristall furnace.

Within a month Lyakhov and Ryumin had embarked on a regime of increased exercise, with sessions of over 2½hr a day, in a new attempt to solve the problems of long-term weightlessness. It was a "150% struggle against weightlessness," according to a Soviet doctor, and it worked. When the cosmonauts came back they were found to be in remarkably good shape. They also pioneered a number of other comforts for the long-term spaceman, taking their first shower on March 23 and subsequently giving themselves a daily wipedown with disposable damp cloths. They had non-foaming toothpaste and shaved using electric razors with suction pipes.

By the end of March Lyakhov and Ryumin had completed 38 repair jobs in addition to their scientific work, including the purging of damaged fuel lines, and all was ready for the first visitors. Unfortunately for them, and for Soyuz 33 crew Rukavishnikov and Ivanov, this craft failed to dock on April 11. The problem caused the ultra-cautious Soviets to cancel the planned visits by a Hungarian and Cuban and to rely on restocking the Salyut with unmanned vehicles. Progress 6, launched on May 13, delivered over 100 items and re-boosted Salyut 6 into a new orbit.

As a result of the cancellation of further manned visits, which meant that Soyuz 32 would have exceeded its 90-day operational life by the time the crew were due to return to Earth, the Russians launched Soyuz 34 unmanned to replace it. The craft, launched on June 6, docked at the aft port and Soyuz 32 was discarded from the front on June 13. Soyuz 34 was then manned by Lyakhov and Ryumin, undocked, backed away while Salyut 6 turned automatically through 180°, and then redocked at the front port, thus making possible further visits by Progress craft.

Progress 7, launched on June 28, delivered the first radio telescope to enter Earth orbit. The cosmonauts then underwent thorough training in its use, watching an instructional videotape that had accompanied the system on Progress 7. The telescope, with its huge dish aerial, was then deployed via the aft port. Measuring 33ft in diameter, the umbrella-like aerial was the largest of its kind, beating by 3ft the unit fitted to the American ATS-6 satellite. Designated KRT-10, the telescope was designed to receive emissions from pulsars and other celestial objects, including the interesting Cassiopeia A. By this time the crew had clocked up a record 140 days in space and Ryumin had, for the first time for a spaceman, put on weight. Their heavy experimental schedule included Earth observations, biology, materials processing and astronomy.

In order to free the aft docking port for new Progress visits it was necessary to discard the radio telescope. This should have been done on August 9 but the telescope jammed against the Salyut's appendages. A spacewalk was urged by the cosmonauts but cautious ground controllers looked at every other possibility, including "shaking" the space station, before finally giving the go-ahead. On August 15 Lyakhov and Ruymin made a hazardous 1hr 23min spacewalk not unlike the Skylab repair mission. Using wire-cutters, Ryumin freed the telescope.

Four days later the cosmonauts came home after 175 days in space. Soyuz 34 landed at 3.30 pm on August 19, touching

Ryumin and Lyakhov pose on the Soyuz/Salyut ground trainer before their flight (Tass)

down 105 miles south-east of Dzhezkazgan. The crew were carried from the capsule and placed in hooded, reclining chairs designed to help the cosmonauts' cardiovascular systems to provide adequate blood flow to their upper extremities. The two spacemen had trouble speaking and were far from comfortable for days: blankets felt like chainmail, soft beds like boards. But they soon recovered, proving the value of their daily exercise routine.

In the most successful Soviet mission ever flown, Lyakhov and Ryumin had taken Russian manned spaceflight hours to a total of 35,775 compared with the US figure of 22,493.

Soyuz 33   April 10, 1979   Flight 84

Name: Soyuz 33
Sequence: 84th astro-flight, 71st spaceflight, 68th Earth orbit
Launch date: April 10, 1979
Launch site: Tyuratam, USSR
Launch vehicle: A2 (SL-4)
Flight type: Ferry to Earth-orbital space station
Flight time: 1 day 23hr 1min 6sec
Spacecraft weight: About 13,230lb
Crew: Nikolai Nikolayevich Rukavishnikov, 46, commander
Maj Georgi Ivan Ivanov, 38, Bulgarian Air Force, cosmonaut researcher

Bulgarian Air Force Maj Georgi Ivanov and the first civilian Soviet commander, Nikolai Rukavishnikov, after the failure of Soyuz 33 to dock with Salyut 6. Ivanov was the 100th man above 50 miles (Tass)

When Soyuz 33 commander Nikolai Rukavishnikov, the first civilian, "non-pilot" commander of a spaceflight, arrived with Bulgarian Georgi Ivanov at the launch pad on April 10, 1979, he said that he was waiting with "great emotion for the launch and docking". Soyuz 33 certainly generated great emotion, though not of the kind envisaged by Rukavishnikov.

Wind speeds approaching 11mph when the cosmonauts went aboard Soyuz 33 had risen to 25mph at the time of launch (8.34 pm), and conditions were described as the worst ever for a Soviet launch. But Ivanov remained calm, his heartbeat peaking at only 74. He was later described as "the coolest cosmonaut in history".

The first manoeuvres by a lighter than usual Soyuz 33 seemed to go as planned. But the cosmonauts were aware that "something" was not quite right, with the main propulsion system combustion chamber pressures reading lower than normal. The main engine was due to fire for six seconds during the approach to Salyut 6 and the eagerly awaiting Lyakhov and Ryumin. It fired erratically for three seconds, then shut down to the accompaniment of abnormal vibrations which the crew felt clearly. An emergency in space had begun, for the incident had directly affected the crew's ability to come home.

The docking cancelled, the two bitterly disappointed cosmonauts powered down and drifted on for a further ten orbits while plans were made for the emergency retrofire and landing. Meanwhile, in Bulgaria nobody knew yet that the mission was a failure. There were mass rallies, interviews with Ivanov's mother and father, and celebrations at Bulgaria becoming the "sixth spacefaring nation".

After a pep talk from cosmonaut chief Shatalov, during which he reminded Ivanov to tuck his moustache into his helmet for re-entry, the nervous cosmonauts made preparations for a single 213sec retro burn. Normally re-entry was carried in two phases to lessen the load on the craft and enable it to make a controlled, lifting re-entry. But Soyuz 33 was returning from a higher orbit than usual, calling for a highly stressed ballistic re-entry subjecting the cosmonauts to up to 10g for 530sec. Though ultimately uneventful, the re-entry was noisy and colourful, with Rukavishnikov likening it to being inside the flame of a blowtorch. Still red-hot, the re-entry module was seen glowing in the darkening sky as it descended towards the landing zone 199 miles south-east of Dzhezkazgan and touched down in a cloud of dust.

Ivanov thus became the only Intercosmos cosmonaut to date not to reach Salyut 6. Born on July 2, 1940, in Lovech, Bulgaria, he took up parachuting as a hobby and became a glider pilot in 1957. He joined the Bulgarian Air Force in 1964 and began training with the Intercosmos cosmonauts in March 1978. Yuri Romanenko and the second Bulgarian, Alexander Alexandrov, were the Soyuz 33 back-up crew.

For Rukavishnikov it was another disappointment in a chequered three-flight space career. He failed to board Salyut 1 during Soyuz 10 and flew a rather quiet, unambitious Soyuz 16. Of Soyuz 33 the sour-faced cosmonaut said: "Machinery is only machinery; things can go wrong."

Soyuz 35   April 9, 1980   Flight 85

Name: Soyuz 35
Sequence: 85th astro-flight, 72nd spaceflight, 69th Earth orbit
Launch date: April 9, 1980
Launch site: Tyuratam, USSR
Launch vehicle: A2 (SL-4)
Flight type: Ferry to Earth-orbital space station
Flight time: 184 days 20hr 11min 35sec (landed in Soyuz 37)
Spacecraft weight: About 14,450lb
Crew: Lt-Col Leonid Ivanovich Popov, 34, Soviet Air Force, commander
Valeri Viktorovich Ryumin, 40, flight engineer

After the return of Lyakhov and Ryumin in August 1979 Salyut 6 remained unmanned until April the following year. But before then the space station had received two unmanned visitors, Soyuz T-1 and Progress 8. Soyuz T-1 was an improved ferry vehicle with an autonomous computer enabling it to make completely automatic rendezvous and dockings, solar panels to extend its endurance, and a new propulsion system. It was launched on December 16, 1979, and stayed docked for only a couple of days. Progress 8, launched on March 27, 1980, boosted Salyut 6 into a higher orbit in preparation for the next marathon visit.

Preparing for this were Soyuz 32 back-up Leonid Popov and Valentin Lebedev, who had flown on Soyuz 13. Popov was born in Alexandria in the Ukraine on August 31, 1945, and joined the cosmonaut corps in 1970. Shortly before Soyuz 35 Lebedev badly injured his knee while exercising on a trampoline and was dropped from the flight. Because the mission was to be of long duration, an experienced flight engineer was needed. Valeri Ryumin, with 175 days in space already under his belt, got the job.

So, just seven months after leaving Salyut 6, he was back in

*Soyuz 35 space duration record-holders Leonid Popov **upper** and Valeri Ryumin (Novosti)*

The Soyuz/Salyut/Progress configuration (Novosti)

the space station again. One of his first actions on entering Salyut 6 was to open a letter that he had written to the next crewman before leaving at the end of his last mission. "I am not in the habit of writing letters to myself," he remarked. The cosmonauts then went to work unpacking Progress 8, and at first the eager Popov rushed around so much that Ryumin had to tell him to slow down. No sooner had Progress 8 been

Ryumin and Popov a day after their return (Novosti)

discarded on April 25 than Progress 9 replaced it on April 27, bringing up a new motor for the biogravity centrifuge and, for the first time in space, transferring water. It undocked on May 20.

Ryumin was plainly enjoying his second stay in Salyut 6. During a televised conference with ground control he produced a huge cucumber. It had grown, he said, with no help from him in a special area of the space station. The scientists were astounded – until Ryumin told them that the marvellous vegetable was made of plastic.

Over the next few months Popov and Ryumin played hosts to cosmonauts from Hungary, Vietnam and Cuba and their Soviet commanders, the Soyuz T-2 crew, and Progress 10 and 11. After landing they said that they had spent 25 per cent of their time receiving visits and unloading cargo. The order of launches was as follows: Soyuz 36 on May 26, Soyuz T-2 on June 5, Progress 10 on June 29, Soyuz 37 on July 23, Soyuz 38 on September 18 and Progress 11 on September 28. Progress 11 stayed docked with Salyut 6 for a record 70 days.

During the August hiatus the crew carried out an important agricultural Earth-resources experiment during which they photographed about 60 million square miles of Soviet territory using the MKF-6 camera, which took 3,500 photographs, and KT-140, which took 1,000.

On October 11 Ryumin and Popov, respectively 10lb and 7lb heavier thanks to a "well ordered life style," boarded Soyuz 37 and landed 112 miles south-east of Dzhezkazgan at 12.50 pm. They seemed well and readapted normally to one g after their record 185-day mission. They had carried out hundreds of experiments and repaired a great deal of equipment, including, to the astonishment of technicians on the ground, the radio telescope dismantled during Ryumin's first stay in the station.

Valeri Ryumin had travelled 150 million miles in nearly a year (362 days). "If it was needed to prove we could go to Mars," he said, "then Leonid and I would volunteer right now."

Soyuz 36　　　　May 26, 1980　　　　Flight 86

Name: Soyuz 36
Sequence: 86th astro-flight, 73rd spaceflight, 70th Earth orbit
Launch date: May 26, 1980
Launch site: Tyuratam, USSR
Launch vehicle: A2 (SL-4)
Flight type: Ferry to Earth-orbital space station
Flight time: 7 days 20hr 45min 44sec (landed in Soyuz 35)
Spacecraft weight: About 14,450lb
Crew: Valeri Nikolayevich Kubasov, 45, commander
Lt-Col Bertalan Farkas, 30, Hungarian Air Force, cosmonaut
researcher

Hungarian Bertalan Farkas signs the Soyuz descent craft with a piece of chalk while mission commander Valeri Kubasov looks on (Novosti)

The next Intercosmos visitor to Salyut 6 was Hungarian Air Force Lt-Col Bertalan Farkas. Born on August 2, 1949, in Gyulahaza in north-east Hungary, he trained as a glider pilot before entering aeronautical engineering college and graduating in 1969. He joined the Army and then the Air Force, and arrived at Tyuratam with colleague Maj Bela Magyari on March 20, 1978, to begin cosmonaut training. Farkas was teamed with Valeri Kubasov and Magyari with Vladimir Dzhanibekov, and early in 1980 Farkas was named as Hungary's first cosmonaut.

Kubasov and Farkas had been aboard Soyuz 36 for 2½hr before their launcher lifted off from Tyuratam at 6.20 pm on May 26. Farkas arrived in Salyut 6 bearing a Hungarian meal for his hosts – goulash, pâté de foie gras, fried pork and jellied tongue – and an array of experiments designated Diagnost, Balaton, Interferon, Dose, Opros, Andio, Oxymeter, Biosphere, Refraction, Zarya, Bealuca, Eotvos and Ilkeminator.

The cosmonauts manufactured gallium arsenide crystals with chromium; studied the misalignment of instruments resulting from prolonged exposure to space; made a photographic geomorphological map of the Carpathian Basin; and studied hearing and human motor response and the possibility of producing the cancer-fighting drug Interferon in zero g.

The routine eight-day mission was quietly received everywhere in the world – except of course in Hungary. It ended with the crew separating from Salyut 6 in Soyuz 35 and firing its retros for 179sec over the South Atlantic. The spacecraft's components separated over Sudan and after a safe re-entry the descent module gently touched down on Russian soil 87 miles south-east of Dzkezkazgan at 6.07 pm on June 3.

Soyuz T-2　　　　June 5, 1980　　　　Flight 87

Name: Soyuz T-2
Sequence: 87th astro-flight, 74th spaceflight, 71st Earth orbit
Launch date: June 5, 1980
Launch site: Tyuratam, USSR
Launch vehicle: A2 (SL-4)
Flight type: Ferry to Earth-orbital space station
Flight time: 3 days 22hr 19min 30sec
Spacecraft weight: About 15,430lb
Crew: Lt-Col Yuri Vasilyevich Malyshev, 38, Soviet Air Force, commander
Vladimir Viktorovich Aksyonov, 45, flight engineer

*Inside Soyuz T-2 are Yuri Malyshev **foreground** and Vladimir Aksyonov* (Novosti)

Ironically, as Soyuz T-2 and cosmonauts Malyshev and Aksyonov made an automatic approach towards Salyut 6 with the aim of completing a "hands-off" docking, the avionics apparently failed and Malyshev had to take manual

Aksyonov in new-style spacesuit (Novosti)

Soyuz T-2 launch. Note the longer launch-escape tower (Novosti)

control at 700ft out. As it happened, the failure indication proved to be spurious.

The new Soyuz, equipped for a crew of three and fitted with new computers, controls, telemetry and solar panels, was launched at 5.19 pm on June 5, just two days after Kubasov and Farkas came home. The launch was marked by two more innovations: Malyshev and Aksyonov wore new spacesuits, and the launcher had an elongated escape tower.

Yuri Malyshev was back-up commander of Soyuz 22, Aksyonov's first mission. Malyshev was born on August 27, 1941, in Nikolayevsk and attended the Kharkov Higher Air Force School, graduating in 1963 and joining the cosmonauts in 1967.

After a briefer than usual stay with Salyut 6 hosts Popov and Ryumin the Soyuz T-2 crew undocked and, for the first time on a Soyuz flight, jettisoned the orbital module and docking system before retrofire. Adopted for the Soyuz Ts, this new procedure saved fuel and demonstrated that the orbital module could be left docked to a Salyut to enlarge its capacity.

Malyshev and Aksyonov landed 124 miles from Dzhez-kazgan at 3.41 pm on June 9. Black soot had covered the windows of the re-entry module, which had been slowed for touchdown by new, enlarged deceleration rockets.

Name: Soyuz 37
Sequence: 88th astro-flight, 75th spaceflight, 72nd Earth orbit
Launch date: July 23, 1980
Launch site: Tyuratam, USSR
Launch vehicle: A2 (SL-4)
Flight type: Ferry to Earth-orbital space station
Flight time: 7 days 20hr 42min (landed in Soyuz 36)
Spacecraft weight: About 14,450lb
Crew: Col Viktor Vasilyevich Gorbatko, 45, Soviet Air Force, commander
Lt-Col Pham Tuan, 33, Vietnamese Air Force, cosmonaut researcher

*A jubilant Pham Tuan **left** and Viktor Gorbatko talk to the press after their return from space* (Novosti)

The flight of Soyuz 37 created a welter of propaganda that proved particularly unpalatable in the United States. The research pilot was Vietnamese Pham Tuan, "the only man to have shot down an American B-52 during the Vietnam War," and one of his jobs on the flight was to "study the effects on the Vietnamese countryside, plants and forest of the enormous amounts of defoliants and fire bombs dropped during the war". Western sports journalists at the Moscow Olympics were asked to applaud the launch during a press conference. Finally, according to Western sources, the launch of a Vietnamese before a Romanian was both an attempt by the Soviets to gain prestige among Third World nations and a deliberate affront to Romania, which was taking an excessively independent line at the time.

The figure at the centre at the centre of this political showpiece was born at Quoc Tuan, North Vietnam, on February 14, 1947. He attended communist youth school and joined the Air Force in 1965, graduating in 1968. He flew in combat during the Vietnam War, though his claim to have downed a B-52 was strenuously denied by the USA. Tuan went to the Gagarin Air Force Academy in 1977, and in 1979 he and Capt Bui Thanh Liem joined the Intercosmos team. Tuan's commander was Viktor Gorbatko and Liem's Valeri Bykovsky.

A spectacular 9.33 pm launch on July 23 was followed by a docking with Salyut 6 at 11.02 pm the following day. Tuan suffered a headache and loss of appetite to start with but soon settled down to work on 30 experiments with Gorbatko and the resident crew of Popov and Ryumin. The major tasks were Earth-resources photography of Vietnam with the KT-140 and MKF-6M cameras. Mission scientists were particularly interested in tidal flooding and erosion, silting of river mouths, hydrological features and the "effects of defoliants".

Tuan and Gorbatko landed in Soyuz 36 at 6.15 pm on July 31, touching down 112 miles south-east of Dzhezkazgan after a flight of nearly eight days.

Name: Soyuz 38
Sequence: 89th astro-flight, 76th spaceflight, 73rd Earth orbit
Launch date: September 18, 1980
Launch site: Tyuratam, USSR
Launch vehicle: A2 (SL-4)
Flight type: Ferry to Earth-orbital space station
Flight time: 7 days 20hr 43min 24sec
Spacecraft weight: About 14,450lb
Crew: Col Yuri Viktorovich Romanenko, 36, Soviet Air Force, commander
Col Arnaldo Tamayo Mendez, 38, Cuban Air Force, research pilot

By the autumn of 1980 the Intercosmos missions to Salyut 6 were becoming rather monotonous to even the most avid spaceflight watcher. Soyuz 38 proved to be no exception.

The crew comprised commander Yuri Romanenko and Cuban Air Force pilot Arnaldo Mendez. Born in Guantanamo, Cuba, Mendez was orphaned early in his childhood. His first job, at the age of 13, was as a shoeshine boy. At 17, following the Cuban Revolution, Mendez joined the Association of Young Rebels and two years later the Army. He then trained to fly fighters. Mendez and his Air Force colleague José Armando Falcon went to the Soviet Union for training in 1978. Falcon and Yevgeni Khrunov were the back-up team for this flight.

A Cuban Communist Party and Government delegation were at the Tyuratam Cosmodrome when at 10.11 pm on September 18 Soyuz 38 lifted off. The craft docked with Salyut 6 at 11.49 pm the following day, on Soyuz 38's 17th orbit. Fifteen experiments were carried out during the mission, including a study on behalf of the Cuban sugar industry of the crystallisation of sucrose in zero g. Electrical

Cuban Arnaldo Tamayo Mendez and Soyuz 38 commander Yuri Romanenko (Novosti)

activity in the human brain was monitored, and Mendez wore shoes which loaded his feet as part of the continuing Soviet effort to find ways of preventing muscular deterioration in zero g.

Soyuz 38 came home on September 26, to the most precise landing in Soviet space history. The capsule touched down at 6.45 pm within 1.8 miles of the planned point, 108 miles from Dzhezkazgan.

On September 29 Salyut 6 completed its third year in space, during which men had been on board for 570 days.

Soyuz T-3   November 27, 1980   Flight 90

Name: Soyuz T-3
Sequence: 90th astro-flight, 77th spaceflight, 74th Earth orbit
Launch date: November 27, 1980
Launch site: Tyuratam, USSR
Launch vehicle: A2 (SL-4)
Flight type: Ferry with repair crew to Earth-orbital space station
Flight time: 12 days 19hr 7min 42sec
Spacecraft weight: About 15,430lb
Crew: Lt-Col Leonid Denisovich Kizim, 39, Soviet Air Force, commander
Oleg Grigoryevich Makarov, 47, flight engineer
Gennadi Mikhailovich Strekalov, 40, research cosmonaut

Right to left *Soyuz T-3 cosmonauts Kizim, Makarov and Strekalov. Makarov was the fifth man to make four flights* (Novosti)

View from Salyut 6, showing communications antenna (Novosti)

At 2.18 pm on November 27, 1980, a Soyuz with a difference was launched. For the first time since June 1971 three Russians lifted off together. They were the crew of Soyuz T-3 and their job was to give Salyut 6 a thorough once-over before the planned arrival of another long-duration crew early the following year.

Among the trio of cosmonauts was flight engineer Oleg Makarov, becoming the first Russian to make four space-flights. The commander was Leonid Kizim and the third crew member was researcher Gennadi Strekalov. Kizim was born in Krasny Liman, in the Ukraine, on August 5, 1941, and graduated from military school in 1963. He joined the cosmonauts in 1965. Strekalov was born on October 28, 1940, in Mytischi, near Moscow, and graduated from the Baumann Technical School in 1965. He worked in a spacecraft design bureau where many cosmonaut engineers learned their trade, and it is said that he watched the fabrication of Sputnik 1 while working as an apprentice coppersmith. He became a cosmonaut in 1973.

With Kizim in the centre couch, Makarov to his left and Strekalov on his right, Soyuz T-3 safely docked automatically at the front port of Salyut 6 at 6.45 pm on November 28. During their hectic two weeks aboard the cosmonauts found time to conduct two experiments. One of these, which contributed to the understanding of crystal behaviour in weightlessness, involved the use of a laser-based holographic system to photograph the dissolution of a salt crystal.

The busy repair schedule included a major overhaul of the hydraulic system, calling for the installation of a new hydraulic unit and pumps. They replaced a programming and timing device in the control system, fitted a new transducer to a compressor in the in-orbit refuelling system, and replaced electronic components in the communications system.

After using the main propulsion system of the docked Progress 11 to raise the space station's orbit, the crew discarded the supply craft, entered Soyuz T-3 and sat through an automatic retro-fire and re-entry, landing at 9.26 am 81 miles south-east of Dzhezkazgan on December 10 after a flight of 12 days. Progress 12 was then launched to dock with Salyut 6 in readiness for the arrival of a new crew in the spring.

Soyuz T-4   March 12, 1981   Flight 91

Name: Soyuz T-4
Sequence: 91st astro-flight, 78th spaceflight, 75th Earth orbit
Launch date: March 12, 1981
Launch site: Tyuratam, USSR
Launch vehicle: A2 (SL-4)
Flight type: Ferry to Earth-orbital space station
Flight time: 74 days 17hr 37min 23sec
Spacecraft weight: About 15,430lb
Crew: Col Vladimir Vasilyevich Kovalyonok 39, Soviet Air Force, commander
Viktor Petrovich Savinykh, 41, flight engineer

Just short of the twentieth anniversary of the launching of the first Russian and the first man into space, flight engineer Viktor Savinykh achieved the distinction of becoming the 100th spaceman and the 50th cosmonaut. He flew on Soyuz T-4, fitted with two seats instead of the three of its predecessor. Savinykh's commander was the 17th man to make three spaceflights and the 10th Russian to do so, Vladimir Kovalyonok. The space given over to the third seat in Soyuz T-3 was now occupied by an abundance of extra film and instrumentation.

Savinykh, born on March 7, 1940, was a specialist in aerial photography and cartography, two disciplines which dominated his stay in Salyut 6. The cosmonauts spent the first seven days after docking on March 14 repairing and refurbishing their new home. They replaced components and freed a jammed solar panel. The loss of power from the panel had resulted in such a reduction of temperature in the station that the interior was dripping with dew. The cosmonauts also installed orientation systems and unloaded scientific equipment from Progress 12, which had been launched to the uninhabited Salyut on January 26, 1981. Progress 12 then undocked at 9.14 pm Moscow time on March 19 and flew solo until 7.59 pm on March 21, when it re-entered over the Pacific.

Kovalyonok and Savinykh completed 30 scientific experiments and hosted the Soyuz 39 and 40 Intercosmos crews before landing 78 miles east of Dzhezkazgan on May 26 after 74 days in space. Salyut 6 had received its last human visitors after 3½ years in service. The Soviets then announced that they would make no more manned flights for a year, though there was one highly significant related event. Cosmos 1267, launched on April 25, docked with Salyut 6 on June 19 to test a multiple docking system designed as the basis for a large modular space station. The prototype of the Star space station module, Cosmos 1267 remained docked with Salyut 6 until they were commanded to re-enter in July 1982.

On September 29, 1981, the Russians suddenly announced that manned flights were to begin again, and later reports indicated that cosmonauts were training in the Arctic for emergency returns to Earth from polar-orbital flights.

A Soyuz T launch (Novosti)

Soyuz T-4 commander Vladimir Kovalyonok with **right** *the 100th man in space, Viktor Savinykh* (Novosti)

Soyuz 39   March 22, 1981   Flight 92

Name: Soyuz 39
Sequence: 92nd astro-flight, 79th spaceflight and 76th Earth orbit
Launch date: March 22, 1981
Launch site: Tyuratam, USSR
Launch vehicle: A2 (SL-4)
Flight type: Ferry to Earth-orbital space station
Flight time: 7 days 20hr 42min 3sec
Spacecraft weight: About 14,550lb
Crew: Col Vladimir Alexandrovich Dzhanibekov, 38, Soviet Air Force, commander
Capt Jugderdemidyin Gurragcha, 33, Mongolian People's Army, cosmonaut researcher

Soyuz 39 commander Vladimir Dzhanibekov and, left, his Mongolian crewman, Jugderdemidyin Gurragcha (Mongolian Embassy via Astro Information Service)

If the next Intercosmos spaceman had been a pioneer in the Gagarin mould, his near-unpronounceable name would have caused havoc in many a television and radio newsroom. But his flight proved to be as colourless as its recent predecessors in the Soviet international programme and the newscasters breathed again.

Mongolia's Jugderdemidyin Gurragcha flew into space aboard Soyuz 39 with Vladimir Dzhanibekov on March 22, 1981. Born on December 5, 1947, into a cattle-breeding family, Gurragcha joined the Army in 1966, went to aviation school in the Soviet Union in 1971 and later graduated from the Zhukovsky Air Force Academy. He joined the cosmonauts in April 1978 with Mongolian colleague Maydarjaviyn Ganzorig, who was teamed up with Vladimir Lyakhov.

The flight profile of Soyuz 39 was typical. Lift-off came at 5.59 pm and docking was achieved at 7.28 pm the following day. Seven days later the eighth Intercosmos crew returned to Earth after completing a wide-ranging scientific programme, including an Earth-resources survey of Mongolia's oil, ore and gas deposits, and use of a visual polarising analyser to assess the effects of prolonged exposure to space on the station's portholes. Soyuz 39 landed in fog and drizzle 106 miles south-east of Dzhezkazgan on March 30.

STS-1   April 12, 1981   Flight 93

Name: STS-1 (OV-102 *Columbia* 1)
Sequence: 93rd astro-flight, 80th spaceflight, 77th Earth orbit
Launch date: April 12, 1981
Launch site: Pad 39A, Kennedy Space Centre, Merritt Island, USA
Launch vehicle: SSME/ET/SRB
Flight type: Earth orbit and return to conventional landing
Flight time: 2 days 6hr 20min 52sec
Spacecraft weight: 219,258lb
Crew: John Watts Young Jr, 50, commander
Capt Robert Laurel Crippen, 43, USN, pilot

The day before the first launch attempt, Bob Crippen, left, and John Young discuss the weather (Nasa)

As the news media assembled at Kennedy Space Centre at the beginning of April 1981, some cynical journalists were preparing to add the Space Shuttle to the list of disasters that had marred the recent history of the United States. After delay upon delay, rocket engine explosions and maddening problems with the crucial thermal protection tiles, the world's most advanced spacecraft was finally ready to go. And go it must, for the Space Shuttle had powerful Congressional enemies who questioned the need for a manned launcher and who would seize on a failure as a pretext to campaign for its cancellation. Thus the future of the US manned space programme depended absolutely on a successful first flight. As it turned out, the mission was a fantastic success. Even the cynics were left cheering as for two days the world marvelled once more at American technology.

The Space Shuttle was a logical development that was an unreasonably long time in coming. Though the technology existed the commitment initially did not, and even when in 1972 the Shuttle was at last given the go-ahead the money allocated for its development was small change by Apollo standards. What has resulted, apart from the delays, is a cut-price system that is not wholly reusable.

The Space Transportation System (STS), as the Shuttle is formally known, consists of four major components: the Orbiter spaceplane, the External Tank (ET) and two Solid Rocket Boosters (SRB). At launch the system weighs 4,458,000lb. The Orbiter and the boosters are reusable, the External Tank is not. The Orbiter is designed to fly 100 times before major refurbishment is necessary, the boosters 25 times. Measuring 120ft long and spanning 78ft, the Orbiter can carry a crew of up to seven: two pilot-astronauts, two astronaut mission specialists and three passenger payload specialists. The Orbiter is attached to the 154ft-long External Tank, which holds the liquid oxygen and hydrogen fuel for the Orbiter's trio of main engines. On either side of the

External Tank are mounted the two 149ft-long Solid Rocket Boosters. At lift-off the five engines develop a combined thrust of over six million pounds.

The Orbiter's reusability, the cornerstone of STS economics, depends on protection from the searing heat of re-entry. The nose and leading edges of the wings experience the highest temperatures, up to 1,260°C, and are protected with reinforced carbon-carbon (RCC) tiles. Other parts of the structure do not get quite so hot and are protected by three other types of material: high-temperature reusable surface insulation (HRSI) for up to 704°C, low-temperature RSI for up to 649°C, and flexible reusable coated Nomex felt for up to 371°C.

Although the first Orbiter, *Enterprise*, had been flown on atmospheric glide tests during 1977 by astronauts Fred Haise, Joe Engle, Gordon Fullerton and Richard Truly, the second Orbiter, *Columbia*, formed part of a rocket-spacecraft combination that, uniquely in the US manned programme, was being used to launch men without first being flight-tested. There can be little doubt of the heroism of the STS-1 crew, John Young and Robert Crippen.

Young had already made four flights, including a Moon landing on Apollo 16, and had served as Apollo 17 back-up commander. As the senior astronaut he was a natural choice as commander of this important mission. His co-pilot, an astronaut since 1969, was making his first flight. Bob Crippen was born in Beaumont, Texas, on September 11, 1937, and graduated in aerospace engineering before joining the US Navy and flying from USS *Independence*. He graduated from the Aerospace Research Pilot School and remained there as an instructor before being selected in the second group of astronauts for the Manned Orbiting Laboratory programme. When this was cancelled Crippen was reassigned to Nasa in September 1969 as a Group 7 astronaut. He served on the 56-day Skylab Medical Experiments Altitude Test (SMEAT) and on the support crews for all three Skylab missions and ASTP.

Side view of STS-1 launch (Nasa)

An anti-climactic computer fault caused the STS-1 launch to be cancelled with only minutes to go on April 10 and provided the anti-Shuttle journalists with material for more knocking copy. But then, at 7 am on April 12, 1981, twenty years to the day since the first manned flight, the Space Shuttle main engines (SSME) ignited, followed by the mighty SRBs. Trailing a dense, golden flame, the Shuttle rose quickly, arcing steeply over the Atlantic. At T+2min 11sec the SRBs fell away, landing at T+7min 13sec in the Atlantic, where they were later recovered.

"What a view! What a view!" exclaimed Crippen as he

Columbia *comes home like an airliner* (Nasa via Astro Information Service)

climbed into space after his 12-year wait. Then, in rapid succession, came SSME shutdown, ejection of the External Tank, which re-entered to destruction over a remote ocean area, and the first of four orbital manoeuvring system (OMS) burns to place *Columbia* in its initial orbit. An hour and 52min after launch the first television pictures showed the payload bay doors open, exposing the cavernous interior of the bay. They also revealed that a number of tiles were missing from an OMS pod, a non-critical area. Though the non-specialist press made much of this, John Young said after the flight that in his opinion the non-critical tiles could be removed altogether on later flights to save weight and increase payload. The astronauts, and Nasa for that matter, did have the added reassurance of knowing that the Orbiter's underside had been seen to be intact in photographs taken by a KH-11 reconnaissance satellite.

Young and Crippen spent the next two days putting the Orbiter gently through its paces, demonstrating its basic habitability and controllability. Then, after 36 orbits, came the moment when the Space Transportation System would be put to the ultimate test. With *Columbia* 172 miles above the Indian Ocean the OMS fired to initiate the re-entry. "Nice and easy does it, John. We are all riding with you," said capcom astronaut Joe Allen. His words echoed the thoughts of the world. Falling at high speed back into the Earth's atmosphere, *Columbia* went into communications blackout at T+53hr 51min.

Young afterwards said that he could see the critical leading-edge and nose tile areas from the cockpit. They turned pinkish at 330,000ft and a speed of Mach 24 and then gradually turned orange-red as the maximum temperature approached 2,700° F. "What a way to come to California!" exclaimed Crippen as the blackout ended after a perfect re-entry.

Columbia touched down at 215mph on a dry-lake runway at Edwards Air Force Base, California. Reflecting the ease and perfection of the flight and landing, Young asked jokingly "Do you want me to put it in the hangar?" as the Orbiter coasted 8,993ft to a stop. *Columbia* had made 36.5 orbits and travelled 933,000 miles in space, and weighed 196,500lb on landing.

Even if they hadn't actually "cleared the road to the stars," as Young optimistically declared after the flight, the astronauts of STS-1 had opened a new era in space exploration and exploitation. President Reagan described the mission as a brave adventure – and then cut $604 million from the Nasa budget. As John Young commented later: "Politicians are a strange bunch of critters."

Soyuz 40   May 15, 1981   Flight 94

Name: Soyuz 40
Sequence: 94th astro-flight, 81st spaceflight, 78th Earth orbit
Launch date: May 15, 1981
Launch site: Tyuratam, USSR
Launch vehicle: A2 (SL-4)
Flight type: Ferry to Earth-orbital space station
Flight time: 7 days 20hr 41min 52sec
Spacecraft weight: About 14,550lb
Crew: Col Leonid Ivanovich Popov, 35, Soviet Air Force, commander
Sen Lt Dumitru Prunariu, 28, Romanian Army Air Force, cosmonaut researcher

*Soyuz 40's Dumitru Prunariu of Romania **left** and Leonid Popov (Novosti)*

The last standard Soyuz was launched into orbit at 3.17 pm on May 15, 1981, carrying the ninth Intercosmos crew and the last from a communist nation before a French cosmonaut flew in 1982. The cosmonaut researcher for the mission was Romania's Dumitru Prunariu and his commander was duration record-holder Leonid Popov. (They were in fact the reserve crew, according to official sources.) The youthful-looking Prunariu was born in Brasov on September 27, 1952, and worked in an aircraft factory before entering the Air Force Regiment of the Romanian Army. He joined the cosmonauts in 1978.

This routine mission, lasting the standard eight days, included a study of the Earth's upper atmosphere and changes in its magnetic field. Soyuz 40 landed 140 miles south-east of Dzhezkazgan at 11.58 am on May 22 after a mission lasting seven days 20hr 38min. The Russians then announced that the two-man Soyuz had been used for the last time. It also seemed that Salyut 6 had seen its last visitors, having served as home to 16 cosmonaut crews and received 15 unmanned craft in its 3½ years of operational life. No fewer than 35 dockings were made by manned and unmanned craft with Salyut 6, which was occupied for a total of 676 days. Some 13,000 photographs of the Earth's surface were taken from Salyut 6, and 1,310 astrophysical, technological and biomedical experiments were carried out aboard the station.

Name: STS-2 (OV-102 *Columbia* 2)
Sequence: 95th astro-flight, 82nd spaceflight, 79th Earth orbit
Launch date: November 12, 1981
Launch site: Pad 39A, Kennedy Space Centre, Merritt Island, USA
Launch vehicle: SSME/ET/SRB
Flight type: Earth orbit and return to conventional landing
Flight time: 2 days 6hr 13min 11sec
Spacecraft weight: 230,708lb
Crew: Col Joseph Henry Engle, 49, USAF, commander
Capt Richard Harrison Truly, 44, USN, pilot

4 mission lasting 2min 34sec. Engle was one of eight astronauts – the others were Richard Truly, Fred Haise, Jack Lousma, Gordon Fullerton, Vance Brand, John Young and Bob Crippen – chosen as the Shuttle space test pilot group in March 1978. He served as John Young's back-up on STS-1.

STS-2 pilot Richard Truly was born in Fayette, Mississippi, on November 12, 1937, and gained a degree in aeronautical engineering in 1959. He joined the Navy in 1960 and later studied in the 1964 class of the USAF Aerospace Research

Lift-off of STS-2 eight days later (Nasa)

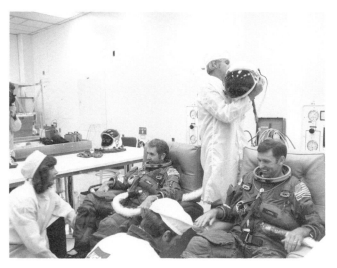

Truly, left, and Engle suit up for the aborted November 4 launch attempt, when the count was stopped at T–31sec (Nasa)

"I would still call it a successful flight even if we had to come home early," said pilot Richard Truly before lifting off in *Columbia* on the STS-2 mission. He and commander Joe Engle did have to come home early, but STS-2 could indeed be considered a success, with 80 per cent of the five-day mission objectives being achieved in just two days.

America's space programme had reached its lowest ebb in the autumn of 1981, despite *Columbia*'s success on the first mission. Some people were even considering shutting off Voyager 2's cameras before its close encounter with Uranus in 1986 in order to save money. So it was in an atmosphere of pessimism and doubt that Engle and Truly prepared to prove that the Space Shuttle really was a reusable system.

For Engle it was the long-awaited climax to an astronaut career that had brought him nothing but frustration since his X-15 astro-flights in the mid-1960s. After serving as Apollo 14 back-up lunar module pilot and being dropped from Apollo 17 to make way for Jack Schmitt, Engle took no part in Skylab or ASTP. When he finally flew again it was on two of the atmospheric Approach and Landing Tests (ALT) with Shuttle Orbiter *Enterprise* in 1977. After separating from the back of a carrier 747 Engle and Truly flew *Enterprise* for 5min 28sec on September 17, 1977. On October 12, 1977, they flew the ALT-

Columbia's thermal protection system has emerged relatively unscathed after the first flight by a recycled spacecraft (Nasa)

Pilot School, becoming an instructor on graduation. In November 1965 he was named among the first group of astronauts assigned to the USAF Manned Orbiting Laboratory programme. This was cancelled in September 1969 and Truly transferred to Nasa. Apart from his ALT experience, he also served on the Skylab and ASTP support crews.

Columbia was scheduled to lift off on September 30, 1981, but did not in fact go until November 12, Truly's 44th birthday. The delays were an embarrassment to Nasa, which was desperately trying to win support for the Shuttle, and resulted initially from the need to replace various items after STS-1. These included a number of thermal tiles, reaction control system (RCS) components, an OMS nozzle and fuel cells. The launch had already slipped to October 9 when, on September 23, nitrogen tetroxide leaked from a ruptured fuel line during propellant loading on the launch pad and spoiled over 240 tiles on the nose. The launch was finally rescheduled for November 4, but when the countdown reached T-31sec the ground launch sequencing computer ordered a hold when it spotted pressurisation problems in the three fuel cells. Attempts to manually override the computer failed and the launch was scrubbed.

This failure proved fortuitious, with subsequent checks revealing contaminants in one of the three auxiliary power units that drove the Orbiter's hydraulic systems. The fault corrected, and with new computer hardware installed, STS-2 at last got under way at 10.10 am on November 12. The flight into orbit went beautifully but then suddenly one of the three fuel cells, being flown for the first time, developed a fault and had to be shut down. Mission rules dictated that an Orbiter down to two operable fuel cells stay in space for no more than two days, so Engle and Truly had to get their skates on if the prime objectives were to be achieved.

Inside the payload bay were the Shuttle's first scientific experiments, packaged as OSTA-1 by Nasa's Office of Space and Terrestrial Applications. The Orbiter was rolled so that OSTA-1 could view the Earth, remotely sensing land resources, environmental quality, ocean conditions, meteorological phenomena and life sciences data. OSTA-1 was mounted on a British-made Spacelab pallet.

OSTA included the Shuttle Imaging Radar (SIR-A) and Measurement of Pollution from Satellites (MAPS) experiments. MAPS charted concentrations of gases in the atmosphere, and data collected on one pass revealed that the concentration of carbon monoxide in the middle troposphere varied from 70 parts per billion over the Americas to 140 ppb over the Mediterranean.

During one pass over the Sahara SIR-A penetrated the Selina Sand Sheet to reveal unsuspected river channels, significant geological structures and possible Stone Age settlements.

Also inside the payload bay was the Remote Manipulator System (RMS), built in Canada. On later flights this will be used to deploy satellites in orbit and to retrieve them for repair, refurbishment or refuelling. The RMS, controlled from the flight cabin, is 50ft long and has a "shoulder," "elbow," "wrist" and "hand". It was successfully tried out during STS-2, resulting in some spectacular television pictures. There were disappointments as well, however: Engle didn't get a chance to test the new Shuttle EVA suit, and didn't want to hear it confirmed that the flight was to be a "minimum mission" of two days, feigning hearing difficulties when the capcom, mission specialist Sally Ride, told him the news.

During re-entry the astronauts flew *Columbia* through a series of aerodynamic response tests designed to evaluate stability and control system effectiveness, and made an automatic descent into Edwards Air Force Base under the guidance of the microwave scanning-beam landing system. Engle took over shortly before touchdown, flew a flare manoeuvre and set *Columbia* down just eight minutes short of the STS-1 flight time. "Welcome home," came a message from one of the three chase aircraft. "Thanks, chase," said Engle matter-of-factly, just as he used to at the end of his routine X-15 missions. Nasa wants the Shuttle to be routine too and STS-2 went a long way towards achieving that aim. The flight had one other distinction: it was probably the last Shuttle mission to enjoy live television coverage at lift-off and on landing.

STS-3   March 22, 1982   Flight 96

Name: STS-3 (OV-102 *Columbia* 3)
Sequence: 96th astro-flight, 83rd spaceflight, 80th Earth orbit
Launch date: March 22, 1982
Launch site: Pad 39A, Kennedy Space Centre, Merritt Island, USA
Launch vehicle: SSME/ET/SRB
Flight type: Earth orbit and return to conventional landing
Flight time: 8 days 0hr 4min 46sec
Spacecraft weight: 235,425lb
Crew: Col Jack Robert Lousma, 46, USMC, commander
Col Charles Gordon Fullerton, 45, USAF, pilot

Fullerton, left, and Lousma at the Shuttle launch pad on February 19, 1982, following a simulated countdown and lift-off and emergency slidewire exercise (Nasa)

Television cameras that wouldn't work, a smeared windshield, unreliable radios, lost tiles, a clogged toilet, a frozen payload bay door latch, and an APU failure: these are the minor problems that the press gleefully highlighted during STS-3. The reporters ignored the fact that these glitches had little effect on the mission, that this was a test flight, and that STS-3 exceeded its operational time in space by a day and yielded scientific riches from a payload bay packed with equipment.

It is a pity that the public was treated to this negative view of the Shuttle, because the programme was coming under enough criticism as it was. Doubts about its ability to launch every two weeks and to be cheaper than expendable launch vehicles were being bandied about, and the £5,450 million programme cost had led to the Shuttle being called a "financial monstrosity" by one leading politician. The Space Shuttle was still fighting for its life in spite of the success of STS-1 and 2, and a catastrophic launch could have meant the end. So there was still plenty to prove on March 22, 1982, when *Columbia* awaited its third lift-off. For the first time the Orbiter was mated to an External Tank painted only in unsightly brown primer: a layer of insulation and paint had been left off, saving 600lb and £7,500.

A minor problem with an erroneous signal from a transponder in a liquid nitrogen purge system delayed the launch by an hour and obscured the fact that the Shuttle was being launched on the very day scheduled when STS-2 had returned 114 days earlier. But then, watched by the biggest crowds since the Moon shots, *Columbia* rose into murky skies at 11 am. A side view on television revealed a perceptible sideways shift by the 4,478,787lb combination as it rose towards a roll and pitch manoeuvre at T+8sec and 400ft. This set the Shuttle up for a more advanced ascent profile, including throttling of the SSMEs from 100 to 68 per cent at the period of maximum dynamic pressure.

A "real barn-burner" was how mission commander Jack Lousma described the launch. After his 59-day flight aboard Skylab 3, Lousma served as back-up to ASTP docking module pilot Deke Slayton. In 1978 he was named as pilot to Fred Haise on the third Shuttle flight, on which it was planned either to boost the Skylab space station into a safe orbit or to cause to re-enter over an area where its debris would present no hazard. But the Shuttle was delayed, Haise resigned and

Taken through the aft window of Columbia's flight deck, this photograph shows the rear of the cargo bay, the Orbiter's fin and Orbital Manoeuvring System pods, and **upper right** *the Plasma Diagnostics Package (PDP) in the grip of the Remote Manipulator System. PDP was designed to study the Orbiter's interaction with its space environment* (Nasa)

Lift-off of STS-3, the first Shuttle mission to incorporate an External Tank painted only with a single coat of primer. Deletion of an insulation layer and a second coat of paint resulted in a 600lb weight saving (Nasa)

Skylab made a spectacular but harmless re-entry in 1979. Lousma was then teamed as commander with Gordon Fullerton, another member of the eight-man Shuttle flight-test team.

Charles Gordon Fullerton was born in Rochester, New York, on October 1, 1936, and received degrees in science and mechancial engineering from the California Institute of Technology in 1957 and 1958. He then joined the USAF from Hughes Aircraft, winning his wings in 1959 and eventually becoming a B-47 bomber pilot with Strategic Air Command. Fullerton graduated from the Aerospace Research Pilot School at Edwards AFB in 1965 and was later selected in the second group of astronauts for the MOL programme. When the military programme was cancelled Fullerton joined Nasa as a Group 7 astronaut and served on the support crews for Apollos 14 and 17. With Fred Haise, Fullerton was the first man to fly the Shuttle Orbiter in the atmosphere, piloting *Enterprise* on the first, third and fifth Approach and Landing Tests in 1977.

As the third of four test flights to prove the system before it went operational, STS-3 had four main objectives: the RMS was to be used to remove heavy payloads from the payload bay, simulating satellite deployment and retrieval; the Orbiter was to be orientated to expose parts of the craft to temperatures as low as –66° C and as high as +93° C for long periods; *Columbia* was to remain in orbit for seven days, during which 14 scientific experiments would be carried out; and the mission was to be "dull and routine," with as little publicity as possible because media attention generally meant problems.

The 10,000lb OSS-1 payload, designed by the Office of Space Sciences, comprised a number of experiments, including the British University of Kent's Microabrasion Foil Experiment, a study of micrometeorites. Experiments not part of OSS-1 included the first Space Shuttle Student Involvement Project package. Todd Nelson, an 18-year-old from Minnesota, had his "Insects in Flight Motion Study" experiment on board.

Columbia duly entered a 130-mile circular orbit, after a minor scare when an overheating APU had to be shut down during launch. Jack Lousma then suffered a recurrence of

Columbia *touches down at White Sands at the end of STS-3* (Nasa)

the motion sickness he had experienced during Skylab 3. But then both astronauts went about things with so much gusto that the schedule was changed to give them more rest.

The failure of two television cameras on the RMS meant that the lifting of one of the scientific packages, the 900lb Induced Environmental Contamination Monitor, had to be cancelled because the pilots could not get a good enough view of the operation on their monitors in the cockpit. The lighter Plasma Diagnostics Package, weighing 353lb, was however successfully manipulated.

The hot and cold soaks of the Orbiter, lasting as long as 80hr in one case, proved that the craft could tolerate the temperature extremes of space. After one cold soak, however, one payload bay door could not be closed because of a frozen latch. The problem was solved by turning the payload bay towards the sun and defrosting the latch.

Missing tiles on the top of the Orbiter's nose and at the rear, 37 in all, bore testimony to high vibration levels and dynamic pressures during the launch. Ground crews found some of the titles on the beach of Cape Canaveral, and Lousma reported that in the early stages of the launch it was like flying through a blizzard, as ice or insulation or both flew off the ET and, it is believed, knocked off a number of the titles.

Evidence that the Soviet Union was "bugging" the mission came when, as *Columbia* flew over a part of the USSR, the astronauts heard a high-pitched whistling in their headsets. This was particularly disturbing to Lousma, who was trying to sleep. A communications fault of another kind took the form of a radio failure. This was in fact caused by an open circuit-breaker and an out-of-position switch, and nothing was actually wrong. The last significant malfunction probably affected the astronauts' comfort most of all: once again, the complex zero-g toilet failed to work properly.

At the end of an exhausting, frustrating seven days in space the crew of *Columbia* prepared for retro-fire and a descent to White Sands, New Mexico, a secondary landing site chosen because the runway at Edwards was under inches of water. But just 40min before retro-fire Lousma and Fullerton were told to "wave off" because the winds were too strong and a dust storm was blowing across the dry runway. This was the second delayed landing in US space history; many flights had come home early but only one other had given the astronauts extra time in space.

A 2min 29sec OMS burn sent *Columbia* home a day later. It was disappointing that White Sands was still to be the landing site and not Cape Canaveral, where a runway was waiting to be inaugurated by an Orbiter and could have received *Columbia* on STS-3. *Columbia* was on autopilot until 120ft, when the undercarriage came down and Lousma took over to fly the Orbiter on to the gypsum runway at 250mph. The main gear touched firmly and the nose began to lower, only to pitch up again suddenly before finally being eased down on to the runway. This manoeuvre, which alarmed onlookers, came about because Lousma thought he had an excessive nose pitch-down rate and over-corrected on the hand controller. *Columbia* came to a halt after travelling about 13,000ft, coasting to an unbraked stop. The Orbiter was in better condition than on its return from the first two flights and, interestingly enough, the troublesome No 3 APU had worked perfectly.

One or two sour notes were sounded after this solidly productive mission. The change of landing site added a week to the turnround time before the final test flight, and *Columbia* was described as "acting like a used car". But then as one Nasa official said: "If it is, I'd like to own it!"

Name: Soyuz T-5
Sequence: 97th astro-flight, 84th spaceflight, 81st Earth orbit
Launch date: May 13, 1982
Launch site: Tyuratam, USSR
Launch vehicle: A2 (SL-4)
Flight type: Ferry to Earth-orbital space station
Flight time: 211 days 9hr 4min 32sec (landed in Soyuz T-7)
Spacecraft weight: About 15,430lb
Crew: Lt-Col Anatoli Berezovoi, 40, Soviet Air Force, commander
Valentin Vitalyevich Lebedev, 40, flight engineer

With the Salyut 6/Cosmos 1267 combination spending its final weeks in space, Salyut 7 was sent into orbit on April 19, 1982. The arrival of the long-awaited replacement for the illustrious Salyut 6 signalled the resumption of Soviet manned flights after a year-long hiatus.

Salyut 7 is not dissimilar to Salyut 6 but its interior is much improved, as described by its first occupant, Anatoli Berezovoi: "At first glance the differences from Salyut 6 do not show. They resemble each other outwardly. Inside is different. It is well lit, comfortable and cosy, with wall panelling painted in bright colours, and it is easy to maintain the workable panels in spotless cleanliness." New equipment include a refrigerator and the Rodkin system, which supplies hot and cold running water at all times. Salyut 7 also has an air-conditioning system which absorbs dust and operates more quietly than earlier installations.

Perhaps the most significant feature of Salyut 7 is its introduction of reinforced docking facilities capable of accommodating vehicles heavier than the 15,500lb Soyuz and Progress. One such is the Star space station module, which as Cosmos 1267 carried out a test docking with Salyut 6. The two ports are equipped with T-bar sights designed to facilitate docking manoeuvres.

Scientific equipment initially fitted included X-ray systems, though the ultra-violet telescope flown on Salyut 6 was not on board. Further equipment was to arrive on Progress ferry vehicles, the first of which, Progress 13, was launched at 9.57 am Moscow time on May 23. Awaiting its arrival were the Soyuz T-5 crew of Anatoli Berezovoi and Valentin Lebedev, who had been on board Salyut 7 since 3.36 pm Moscow time on May 14, following a launch from Tyuratam at 1.58 pm the previous day. Under what was now a standard practice, Soyuz T-5's third crew position had been occupied by 220lb of extra equipment.

Lt-Col Anatoli Brezovoi, a space first-timer, was the commander. Born on April 11, 1942, in the Oktyabrsky district of Russia, he worked as a lathe operator in Novocherkassk before graduating from Kachnish Military High School for Test Pilots in 1965. He served as an Air Force pilot before joining the cosmonauts in 1970. Lebedev had recovered from the leg injury that had put him out of the running for the long-duration Soyuz 35/Salyut 6 flight.

After mothballing Soyuz T-5 and settling into the 20° C and 840mm Hg environment of their plush space hotel, the cosmonauts turned on the Stroka teleprinter and the Delta navigation system and got down to work. Doctors claimed that Lebedev's body "remembered" zero g and that this helped ease his readaptation to it. He reacted to weightless-

Soyuz T-5 commander Anatoli Berezovoi (Novosti)

ness "as if it were yesterday", they reported.

On May 20 the cosmonauts performed a space first by hand-launching the Iskra 2 satellite from the airlock of Salyut 7. This satellite, a radio enthusiasts' comsat similar to the American Oscars, was the first to be deployed from a space station, although Apollo command modules had launched sub-satellites in lunar orbit and the Space Shuttle would soon be launching full-scale comsats in Earth orbit.

When Progress 13 docked at Salyut 7's service module at 11.57 am Moscow time on May 25 the space station complex increased in length to 95ft. Progress 13 carried 660kg of fuel for the station's integrated propulsion system and 290lit of water, which was transferred into spherical vessels on the exterior of Salyut 7 rather than into compartments inside.

Apart from the X-ray installation, scientific equipment on the space station included a Kristall materials-processing furnace, an EFO-7 electrophotometer for star studies, and the Oasis plant-growing experiment. Previous attempts at plant cultivation had not proved successful and it was hoped that this system, with improvements to the watering and root aeration arrangements, would yield a useful crop of peas and onions.

The engine of Progress 13 was fired in orbital manoeuvres

carried out to prepare Salyut 7 to receive the international crew of Soyuz T-6. The unmanned ferry then undocked at 10.31 am Moscow time on June 4, and French cosmonaut visiting crew had left Salyut 7 in Soyuz T-5, leaving the fresh craft for the long-stay cosmonauts, it was expected that Berezovoi and Lebedev would be receiving more visitors. But

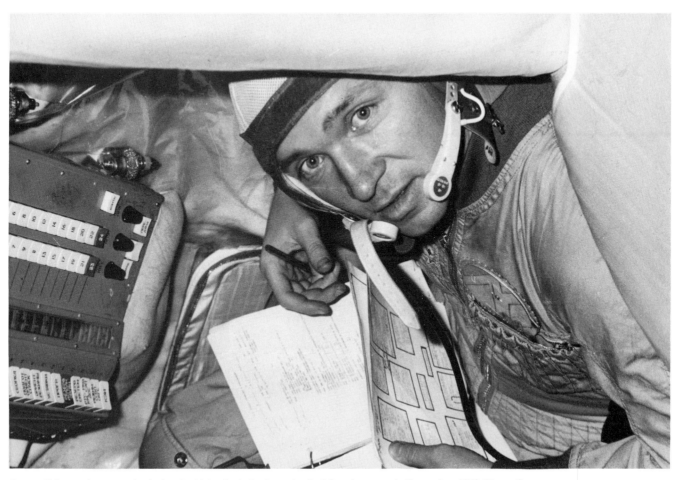

Soyuz T-5 was the second mission for Valentin Lebedev, who first flew in space in December 1973 (Novosti)

Jean-Loup Chrétien and his two Russian companions began their visit on June 25. A second Progress, No 14, arrived on July 12 with bulk cargo, including more water and fuel.

With over 600 agencies in the Soviet Union alone making use of Earth-resources photographs from the Salyut/Soyuz programme, there was bound to be plenty of work for the MKF-6M multi-spectral camera. Its first targets were the winter grain crops in Krasnodar, the principal granary. Berezovoi and Lebedev were reported to have discovered a promising oil and gas field on the left bank of the Volga near Astrakhan, and a lead deposit in Eastern Yukutia.

On July 30 the cosmonauts carried out a spacewalk during which Lebedev went outside to disassemble and partially replace scientific equipment, and to retrieve samples of materials that had been exposed to space since Salyut 7 reached orbit. Berezovoi monitored his activities from the airlock and took live television pictures. Lebedev also tested tools and methods of making mechanical joints. The hatch opened at 6.39 am Moscow time and closed 2hr 33min later. It was the cosmonauts' 78th day in space.

The next mission high point came on August 20 with the arrival of the world's second spacewoman, Svetlana Savitskaya, and her two male colleagues on Soyuz T-7. After the

– Progress 15 and 16 excepted – it was not to be, and so the two men spent a lonely autumn working like beavers, concentrating on Earth-resources and materials-processing experiments in particular. Almost forgotten in the West, they were regular stars on Soviet television, while the long days of confinement were made tolerable by tape-recordings of the sounds of the Earth.

November 18 saw the hand-launching of another Iskra miniature comsat. By early December the astronauts had completed over 60 Earth-observation sessions, taking over 2,000 pictures with the MKF-6M and more than 3,000 with the KT-140 topographical camera.

Progress 16 was used to perform a number of orbital adjustments on December 8, the first hint that Salyut 7's crew were preparing to leave. They wanted to get home for the New Year holidays and had started to show signs of tiredness and irritability, so much so that their sleep period had been extended from eight hours to 12. But because they would need at least two weeks' readaptation under medical supervision after their return, Berezovoi and Lebedev could land no later than mid-December. The inability of Salyut 7 to make radical orbital changes to ensure a daylight landing for Soyuz T-7 in a prime recovery area, and the lack of a Soyuz night retro-fire capability, meant that the only suitable recovery time was late on December 10. This allowed Berezovoi to make a daylight orientation of Soyuz T-7 before retro-fire, and the two cosmonauts landed 118 miles east of Dzhezkazgan at 10.03 pm.

The weather at the landing site – thick fog, heavy snow and an outside temperature of 0° F – was so bad that helicopters could not reach it. To make things worse, the re-entry module landed particularly hard and rolled down a slope, hurling the cosmonauts around. When the spacecraft came to rest, Lebedev was lying on top of Berezovoi. They had to wait 20 minutes before a surface rescue team found them, and because of the weather they spent their first Earth night in seven months in the back of a truck.

The new space duration record-holders returned by helicopter to Tyuratam on December 11 and, looking tired and drawn, talked to the press and television on December 14. They had lost several pounds in weight, their red blood cell counts were reduced, their pulse rates and blood pressures were high, and they were allowed only a 500 ft walk per day for the first two days. On January 13, 1983, it was reported that they had still not fully recovered and were suffering from a "space hangover".

As of mid-April 1983 there was still no new crew in Salyut 7, following the abortive Soyuz T-8. However, a Star module, Cosmos 1443, had completed a successful automatic docking on March 10.

Soyuz T-6 June 24, 1982 Flight 98

Name: Soyuz T-6
Sequence: 98th astro-flight, 85th spaceflight, 82nd Earth orbit
Launch date: June 24, 1982
Launch site: Tyuratam, USSR
Launch vehicle: A2 (SL-4)
Flight type: Ferry to Earth-orbital space station
Flight time: 7 days 21hr 50min 52sec
Spacecraft weight: About 15,430lb
Crew: Lt-Col Vladimir Alexandrovich Dzhanibekov, 40, Soviet Air Force, commander
Alexander Sergeyevich Ivanchenkov, 41, flight engineer
Lt-Col Jean-Loup Chrétien, 44, French Air Force, research pilot

France has always had an active interest in space, running a buoyant programme independently of the European Space Agency, in which the French also play a leading part. France was the third country to launch a satellite and the first Western nation to co-operate with the Soviet Union in space. It was not surprising therefore when in 1980 two French pilots joined the Intercosmos cosmonaut team with a view to one of them being launched aboard a Soyuz two years later. Moreover, at the time the Soviet Union invited France to join the programme, the USA was scheduled to launch the first Western European into space aboard the Shuttle in late 1982 – and the four European astronaut candidates did not include a Frenchman.

As it turned out, the French cosmonaut was the first Western European into space, though there were many who felt that, in the light of events in Afghanistan and Poland, he should not have been allowed to go ahead with the flight. As a result, the French Government resisted its natural inclination to make the most of the mission, emphasising that it was purely scientific in nature. The French spaceman, Jean-Loup Chrétien, was described as a *spationaute* rather than a cosmonaut, and President Mitterand stressed the scientific importance of the mission when he sent a message of goodwill to the crew.

Chrétien, born on August 20, 1938, at La Rochelle, has a distinguished test-flying background, having served as chief test pilot for the Mirage F.1 fighter.

The crew – commander Vladimir Dzhanibekov, flight engineer Alexander Ivanchenkov, and Chrétien – were launched from Tyuratam at 7.39 pm Moscow time on June 24. Their mission was to become the first guests of Salyut 7 residents Berezovoi and Lebedev. The Soyuz T-6 back-up team comprised Leonid Kizim, new cosmonaut Vladimir Solovyov and Frenchman Patrick Baudry. Dzhanibekov was drafted in to replace Yuri Malyshev when the latter was pulled out for personal reasons in January 1982.

During their intensive pre-launch training the two crews drew lots to decide what emergencies they would tackle in the simulator. The prime crew found themselves faced with a failure of the radio control system during rendezvous, necessitating a manual docking. It proved a lucky fall of the dice: shortly before docking with Salyut 7 over Libya at 9.46 pm Moscow time on June 25, the hitherto totally automatic flight – television during launch and ascent showed the cosmonauts sitting impassively with their arms folded – went manual when the spacecraft's computer failed during rendezvous. Three hours later, a Salyut was housing five people for the first time ever, and there were five in space for

The Soyuz T-6 Intercosmos crew. **Left to right** *Frenchman Jean-Loup Chrétien, the first* spationaute *and the first Western European in space; mission commander Vladimir Dzhanibekov; and flight engineer Alexander Ivanchenkov (Novosti)*

the first time since 1975. Nasa's STS-4 Shuttle flight overlapped with this event, bringing the total of men simultaneously in space to seven for the first time since 1969.

The international mission was dominated by medical experiments. One of these was Echograph, a French system which was left on Salyut 7 for use by the long-stay crew. This instrument enabled the cosmonauts to monitor their heart and circulation in real time. The information was displayed on a television monitor and stored in the on-board computer.

At 3.01 pm Moscow time on July 2 Soyuz T-6 left Salyut 7, having completed 125 orbits. Orbital module separation came one orbit later, followed by a 200sec retro burn at 5.35

pm on orbit 127. The service module was jettisoned 11min later, and the re-entry began after another 11min. The main parachutes opened at 6.10 pm and Soyuz T-6 touched down at 6.21 pm in a field near Arkalyk. Chrétien afterwards reported finding the re-entry and landing more dramatic than the launch.

Subsequent reports revealed that one of the quieter missions of the unglamorous Intercosmos series had in fact been livelier than at first appeared. There had been so much work to do that Chrétien and his Soviet colleagues criticised the schedule, while the long-stay residents requested a day off afterwards to recover.

STS-4 June 27, 1982 Flight 99

Name: STS-4 (OV-102 *Columbia* 4)
Sequence: 99th astro-flight, 86th spaceflight, 83rd Earth orbit
Launch date: June 27, 1982
Launch site: Pad 39A, Kennedy Space Centre, Merritt Island, USA
Launch vehicle: SSME/ET/SRB
Flight type: Earth orbit and return to conventional landing
Flight time: 7 days 1hr 9min 31sec
Spacecraft weight: 241,664lb
Crew: Capt Thomas Kenneth Mattingly II, 46, USN, commander
Henry Warren "Hank" Hartsfield, 48, pilot

The astronauts stepped out as though they were going shopping. The man reading the countdown to the press and public at 11 am on June 27 didn't bother to finish. All in all, the launch of STS-4 received scant attention from everyone but the men actually doing the job. The Space Shuttle had succeeded in making spaceflight routine, and it still had one test flight to go before being declared operational. Another

reason for the lack of coverage was the fact that Nasa wasn't allowed to reveal much about DOD-82-1, the first military payload to be carried on a US manned flight. For the first time in 24 years newsmen encountered a change from the Nasa policy of open-handedness to which they had become accustomed. In-flight television was limited because the public could not be allowed to see the top-secret package. On the other hand, the man in the street manifested a lively interest in the mission, with over a million people using a special telephone line to Nasa for flight information. If timed right, these calls even presented an opportunity to listen in on space-ground communications.

The crew were the first US astronauts not to have back-ups, the result of a Nasa decision that there were enough flight-experienced crewmen available to stand in without specific training. They were also the first Shuttle crew to lift off on time. The commander was Thomas Mattingly, the US Navy captain who, as command module pilot of Apollo 16, had orbited the Moon in 1972. Although not originally scheduled to make one of the four test flights, Mattingly won command of STS-4 following his assignments as back-up

Columbia *throws a Concordesque shadow as it flares before touching down on the concrete runway of Edwards Air Force Base at the end of* STS-4 *(Nasa)*

commander of STS-2 and STS-3.

Henry Hartsfield, teamed with Mattingly for STS-4, was one of the ex-MOL astronauts who joined the Nasa Group 7 team in 1969. Born in Birmingham, Alabama, on November 21, 1933, Hartsfield graduated in physics, astronautics and engineering science, gaining the last degree while serving as an astronaut. He joined the US Air Force in 1955 and graduated from the USAF Test Pilot School, serving as an instructor there before joining MOL in 1966. He served on the support crews of Apollo 16 and Skylabs 2, 3 and 4 before resigning from the Air Force in August 1977.

Columbia had arrived back at the Cape from White Sands on April 6 in need of some repair following gypsum dust contamination of fuel pumps; a number of damaged tiles also needed to be replaced. A turnround time of 64 days was achieved.

In addition to scientific and military tasks, the STS-4 flightplan included a tight schedule of manoeuvres designed to expose certain parts of the Orbiter to Sun and shadow for long periods. One of these orientations, belly to Sun, was included in the hope that the heat would evaporate water trapped beneath the tiles during a thunderstorm only days before the launch; in the same downpour a number of tiles were damaged by ¼in hailstones.

It is thought that the water in the tile system increased the launch weight, possibly contributing to an initially depressed launch profile as the Shuttle turned due east toward its 28° orbit. In an effort to make up the shortfall the SSMEs were commanded to burn for 3sec longer, but even after a number of OMS burns the orbit was still 7km short, which meant that *Columbia* would touch down 5min earlier at the end of the mission.

A far worse launch problem was the initially inexplicable loss of both SRBs under 3,000ft of water after a possible parachute failure. A robot underwater sled later returned detailed pictures of the wreckage, revealing that explosive bolts on half of the parachute risers had fired prematurely, causing the parachutes to stream uselessly instead of filling with air.

Hank Hartsfield, left, and Ken Mattingly pictured at the launch pad during their stint as the back-up crew for STS-2 (Nasa)

There were actually two military payloads on board *Columbia*: an ultra-violet horizon scanner and the Cirris cryogenic infra-red radiance instrument. The latter is designed to look closely into the atmosphere to obtain spectral data on the exhausts of vehicles powered by rocket or air-breathing engines, with a view to developing a space-based early-warning system. Cirris needed liquid helium cooled to –270°C to give a low-temperature background so that the sensitive infra-red scanner could acquire high-resolution data. The payload therefore had to be loaded on to *Columbia* as the Orbiter stood on the launch pad only hours before launch.

Mounting criticism of the military nature of STS-4 had already resulted in the dropping of a reconnaissance camera from the flight. USAF Space Command then suffered another blow when Cirris provided no data at all because the lens cap remained jammed on. The astronauts discussed methods of getting the cap loose with the payload communicator at the Air Force Satellite Control Facility at Sunnyvale, California, suggesting that they could knock it off with the RMS or that Mattingly could go EVA to remove it. However, because STS-4 was a test flight and its payload experiments considered a bonus, Nasa management decided to call a halt to the attempts to get Cirris operational.

Science experiments in the mid-deck area of *Columbia* included the Continuous Flow Electrophoresis System, designed to evaluate the use of zero-g platforms for drug manufacture. Some scientists believe that by 1987 drugs that can only be made in space will be on the market. These could include beta cells, a single-injection cure for diabetes; high-yield, high-purity Interferon; epidermal growth factor products for treating burns; hormone products capable of stimulating bone growth; and up to 40 other products now under study. The STS-4 mission contributed significantly to the achievement of this goal.

The first commercial payload carried by the Shuttle, a "Getaway Special," was also on board. This facility for flying small, cheap experiments was privately purchased and donated to students of Utah State University, who devised a combination of experiments in metallurgy, algae growth and fruit fly behaviour. The package took a long time to power up, however, rather limiting its effectiveness. There were also two student-involvement experiments, four other Nasa experiments, including a Latex reactor, and three instrument packages, two to record aerodynamic data and the IECM, an environmental contamination monitor first flown on STS-3. The IECM was successfully carried out of the payload bay by the RMS to "sniff" around the Orbiter for contaminants that might spoil experiments.

The astronauts took time off from their busy schedule to stage in-flight television programmes. During one of these sessions Mattingly indulged in some gentle counter-propaganda. Looking forward to their Independence Day landing, he said: "It's kind of fitting that we land on July 4 and celebrate the ushering in of a new era, just as our forefathers ushered in an era of democracy for the whole world over 200 years ago on the same date". Hartsfield's statement was more to the point, and probably directed at his Congressional masters: "We're going to explore space, we're going to put it to work for ourselves and make it pay off like it never has done. There is a whole frontier out there."

Towards the end of the flight Mattingly tried on the new Shuttle spacesuit in the airlock. The only problem, he said afterwards, "is that I didn't get to open the door." No EVA had been scheduled for STS-4, so Mattingly had to stay put inside the Orbiter.

During re-entry Mattingly performed a pushover flight-test manoeuvre at Mach 13 that resulted in double the anticipated sink rate. The subsequent 20-mile downrange error was automatically corrected by the flight control system.

Amid a welter of Independence Day celebrations *Columbia* settled onto the concrete Runway 22 at Edwards. Waiting to receive the two astronauts was President Reagan, who wasn't quite as enthusiastic about the Shuttle as Hartsfield had been, making a lukewarm speech which did not include the expected intimation that funds for a manned space station were on their way. "We must look aggressively at the future," he said, "by demonstrating the potential of the Shuttle and establishing a more permanent presence in space." Still, it was better than nothing to a funds-hungry Nasa.

As *Columbia* was prepared for the first operational mission, STS-5, a newspaper advertisement placed by Shuttle main contractor Rockwell summed up the achievement of the programme so far, declaring: " The Doors To Space are Open." And so they were.

Soyuz T-7   August 19, 1982   Flight 100

Name: Soyuz T-7
Sequence: 100th astro-flight, 87th spaceflight, 84th Earth orbit
Launch date: August 19, 1982
Launch site: Tyuratam, USSR
Launch vehicle: A2 (SL-4)
Flight type: Ferry to Earth-orbital space station
Flight time: 7 days 21hr 52min 24sec (landed in Soyuz T-5)
Spacecraft weight: About 15,430lb
Crew: Col Leonid Ivanovich Popov, 36, Soviet Air Force, commander
Alexander Alexandrovich Serebrov, 38, flight engineer
Svetlana Yevgenyevna Savitskaya, 34, research cosmonaut

The launch of the second Soviet spacewoman ahead of American Sally Ride, whose selection for a Shuttle mission had just been announced, had an air of familiarity for those who remembered the early days of manned spaceflight. The fact that no woman had followed Valentina Tereshkova for almost 20 years strongly suggested that Soyuz T-7 was every bit as much a propaganda effort as the unhappy Tereshkova's flight had been. It was hard to believe *Pravda*'s assertion that it had taken two decades of research to ensure that life in space was tolerable for women.

The second woman in space was Svetlana Savitskaya, research cosmonaut aboard Soyuz T-7, which was launched at 9.12 pm Moscow time on August 19. Born in Moscow on

*Aboard Soyuz T-7 were **left to right** Svetlana Savitskaya, Leonid Popov and Alexander Serebrov (Novosti)*

Serebrov, Savitskaya and Popov in zero-g training for Soyuz T-7 (Novosti)

August 8, 1948, Savitskaya trained as a pilot, became a flying instructor, won the world aerobatic championship in 1970, and by 1976 was a qualified test pilot. She has test-flown 20 aircraft and holds 18 aviation world records.

The Soviet media went wild over the new spacewoman, and not since the Soyuz 11 tragedy had space taken up so much newspaper space. Awaiting her arrival aboard Salyut 7 were cosmonauts Berezovoi and Lebedev, while she was accompanied by veteran Leonid Popov and flight engineer Alexander Serebrov, also making his first flight. Serebrov was born in Moscow on February 15, 1944, and had worked from 1976 in a spacecraft design bureau before his selection as a cosmonaut.

Soyuz T-7 docked with Salyut 7 at 10.32 pm Moscow time on August 20. Watched by live television cameras, Savitskaya crawled through the airlock and into the space station, to be greeted by a jocular Lebedev. "We've made you a little apron," he said, and then suggested she get to work on the housekeeping. Savitskaya instantly made it clear that she wouldn't be doing any such thing: "Housekeeping details are

the responsibility of the host cosmonauts," she replied firmly.

Savitskaya's main duty during the mission was to operate the life sciences experiments, three of which were designated Braslet, Vorotnik and Koordinatsiya. Braslet imposed an Earth-equivalent load on the cardiovascular system by restricting the blood flow to the upper body. Vorotnik was concerned with the study of motion sickness in space and its prevention, with the related Koordinatsiya looking at the role of eye movement in provoking this troublesome phenomenon. Other experimental equipment included an electrophoresis system in which cells were separated in zero g by an electric current, and a photometer to determine the densities of space-originating aerosols captured in the upper atmosphere.

The Soyuz T-7 crew came home in Soyuz T-5, leaving a fresher craft for the long-duration crew. T-5 landed 70 miles north-east of Arkalyk at 7.04 pm Moscow time. Doctors pronounced Savitkaya fit and well after the flight, which seems to have been as much of a success as Tereshkova's was an ordeal.

Name: STS-5 (OV-102 *Columbia* 5)
Sequence: 101st astro-flight, 88th spaceflight, 85th Earth orbit
Launch date: November 11, 1982
Launch site: Pad 39A, Kennedy Space Centre, Merritt Island, USA
Launch vehicle: SSME/ET/SRB
Flight type: Satellite deployment in Earth orbit
Flight time: 5 days 2hr 14min 26sec
Spacecraft weight: about 241,200 lb
Crew: Vance DeVoe Brand, 51, commander
Col Robert Franklyn Overmyer, 46, USMC, pilot
William Benjamin Lenoir, 43, mission specialist
Joseph Percival Allen, 45, mission specialist

Probably the most significant manned spaceflight since Apollo 11, STS-5 passed off very quietly. The first commercial Shuttle mission, it was highlighted by the deployment of two communications satellites in orbit. Launch took place on time at 7.19 am on November 11, 1982. The crew was the first four-man team in history and the first American crew to fly without spacesuits and with no means of emergency escape. The crew comprised ASTP veteran Vance Brand, ex-MOL astronaut Bob Overmyer, and two Class 7 astro-scientists, Bill Lenoir and Joe Allen. The SBS-C business comsat was successfully deployed on orbit 6 at T+7hr 58min 35sec, followed by Canada's Anik C-3 on orbit 22. Spacesickness forced postponement of the planned EVA by Lenoir and Allen, and when they tried again the following day faulty spacesuits finally put paid to the attempt. Lenoir and Overmyer's nausea was publicised and discussed endlessly, much to the chagrin of the crew. A perfect concrete runway landing was made at Edwards AFB on November 16.

SBS-3 is deployed from Columbia's *payload bay (Nasa)*

Columbia mirrored in a pool left by a rare rainstorm at Edwards AFB following the STS-5 mission (Nasa)

The first four-man crew in space pose for an automatic camera after the successful deployment of two satellites. They are, clockwise from twelve o'clock, Bill Lenoir, Bob Overmyer, Joe Allen and Vance Brand (Nasa)

Name: STS-6 (OV-99 *Challenger* 1)
Specification: 102nd astro-flight, 89th spaceflight, 86th Earth orbit
Launch date: April 4, 1983
Launch site: Pad 39A, Kennedy Space Centre, Merritt Island, USA
Launch vehicle: SSME/ET/SRB
Flight type: Satellite deployment in Earth orbit
Flight time: 5 days 0hr 23min 42sec
Spacecraft weight: 258,529lb
Crew: Paul Joseph Weitz, 50, commander
Col Karol Joseph Bobko, 45, USAF, pilot
Franklin Story Musgrave, 47, mission specialist
Donald Herod Peterson, 49, mission specialist

STS-6 was originally due to fly on January 27, 1983, but a static engine test of the new Orbiter, *Challenger*, in December 1982 revealed a hydrogen leak. An engine change was ordered, but tests of the new powerplant showed up another fault, a potentially disastrous oxygen leak. Eventually, after more engine changes and problems with contamination of the satellite payload, *Challenger* left the launch pad at 1.30 pm local time on April 4, 1983. Although the Orbiter performed near-flawlessly during the five-day flight, the successfully deployed payload, the first Tracking and Data Relay Satellite, placed itself in the wrong orbit. A spacewalk by Story Musgrave and Don Peterson, featured in superb colour television pictures, proved the design of new Shuttle spacesuits and the ability of astronauts to work in the payload bay. *Challenger* performed a textbook re-entry, touching down at Edwards AFB before being returned to the Cape to be prepared for another trip in mid-June.

Above Challenger *rolls and pitches into her ascent trajectory above Pad 39A at the start of STS-6* (Nasa)

Below *Musgrave, left, and Peterson in* Challenger's *payload bay during the first US EVA since 1974* (Nasa)

Left *Most Shuttle crews have an unofficial, lighthearted photo. This one shows Paul Weitz's "F Troop". Standing are Story Musgrave (left) and Karol Bobko, with Weitz seated and Don Peterson kneeling* (Nasa)

Name: Soyuz T-8
Specification: 103rd astro-flight, 90th spaceflight, 87th earth orbit
Launch date: April 20, 1983
Launch site: Tyuratam, USSR
Launch vehicle: A2 (SL-4)
Flight type: Ferry to Earth-orbital space station
Flight time: 2 days 0hr 17min 48sec
Spacecraft weight: Approx 15,430lb
Crew: Lt-Col Vladimir Georgyevich Titov, 36, Soviet Air Force, commander
Gennadi Mikhailovich Strekalov, 43, flight engineer
Alexander Alexandrovich Serebrov, 39, second flight engineer

Left to right: Serebrov, Titov and Strekalov at the Star City simulator (Tass)

Soon after Soyuz T-8 entered orbit the rendezvous radar antenna failed to deploy. Normally the mission would have been aborted but this time the crew were given the go-ahead to try a rendezvous and docking with Salyut 7/Cosmos 1443 using visual reference and radar inputs from the ground. Reports on the flight were deleted from Radio Moscow's news after the 11th orbit, and the crew were aware that their docking run, during orbit 19, had a "low probability of success". Mission commander Titov explained later that he got to within 525ft of Salyut 7 before abandoning the attempt, fearing a catastrophic collision because his approach speed was too high. Soyuz T-8 came home at 5.29pm the same day. Apart from its disappointing conclusion, the flight had another two distinctions: Titov was the second cosmonaut of that name, and Serebrov was the first person ever to fly consecutive missions in a national series.

Name: STS-7 (OV-99 *Challenger* 2)
Sequence: 104th astro-flight, 91st spaceflight, 88th Earth orbit
Launch date: June 18, 1983
Launch site: Pad 39A, Kennedy Space Centre, Merritt Island, USA
Launch vehicle: SSME/ET/SRB
Flight type: Satellite deployment and retrieval
Flight time: 6 days 2hr 23min 59sec
Spacecraft weight: 247,178lb
Crew: Capt Robert Laurel Crippen, 45, USN, commander
Capt Frederick "Rick" Hauck, 42, USN, pilot
Col John McCreary Fabian, 44, USAF, mission specialist
Sally Kristen Ride, 32, mission specialist
Norman Earl Thagard, 39, mission specialist

Above First full picture of the Shuttle in space was taken by a camera on the SPAS free-flyer. Note the RMS cocked to form a figure seven (Nasa via Astro Information Service)

Right The first five-person space crew: left to right, Thagard, Crippen, Hauck, Ride and Fabian (Nasa via Astro Information Service)

Below STS-7 comes home to Edwards AFB after being waved off from Kennedy (Rockwell International)

Nasa played down the fact that the crew of STS-7 included Sally Ride, the first American woman in space, to such an extent that what could have been a highly publicised mission passed quietly. It was the way both Ride and Nasa wanted it. Originally scheduled with a four-person crew, STS-7 became the first in history to fly five people. Dr Norman Thagard had been added to study the spacesickness, or space adaptation syndrome, which was regularly afflicting Shuttle crews and clearly worrying Nasa. *Challenger* was launched on time, with the first Shuttle two-timer, Robert Crippen, in the commander's seat.

Three satellites were deployed: the Palapa and Anik communications satellites, and a unique retrievable free-flier from Germany called SPAS. On this flight it was used by Fabian and Ride to test the ability of the Remote Manipulator System to retrieve objects in orbit, and its on-board cameras succeeded in taking breathtaking views of *Challenger* in orbit against the black of space and the blue of Earth. *Challenger* was due to make the first Shuttle landing at the Kennedy Space Centre, but bad weather first caused a two-orbit extension of the flight and then a cancellation and diversion to Edwards Air Force Base.

Name: Soyuz T-9
Sequence: 105th astro-flight, 92nd spaceflight, 89th Earth orbit
Launch date: June 27, 1983
Launch site: Tyuratam, USSR
Launch vehicle: A2 (SL-4)
Flight type: Ferry to Earth-orbital space station
Flight time: 149 days 10hr 46min
Spacecraft weight: Approx 15,430lb
Crew: Col Vladimir Afanasevich Lyakhov, 42, Soviet Air Force, commander
Alexander Pavlovich Alexandrov, 40, flight engineer

Above *Soyuz T-9 crew Lyakhov (left) and Alexandrov* (Tass)

Below left *Lyakhov during his EVA. Note the multiple solar panels on Salyut 7* (Tass)

Below centre *The launch escape system that saved the lives of the Soyuz T-10-1 crew when their rocket exploded on the pad* (Tass)

Below right *Lyakhov, left, and Alexandrov are welcomed by photographers, reporters and traditional bread and salt after their landing* (Tass)

Soyuz T-9 docked safely at the rear of Salyut 7 on June 28, 1983. Attached to the other end, almost as large as the space station itself, was the second Star Module. Designated Cosmos 1443, this 13m-long tug and electrical power module separated from Salyut 7 on August 14, ejecting a re-entry vehicle which landed on August 23. The major portion was then destroyed in a deliberate re-entry. Soyuz T-9 flew from the rear to the front of Salyut 7 to await the arrival of Progress 17, carrying supplies, mail and fuel, on August 19. On September 9 the space station suffered a serious propellant leak, obliging Lyakhov and Alexandrov to move into the Soyuz for a possible return to Earth. In the end this proved unnecessary and the two cosmonauts worked on, awaiting the arrival of a two-man maintenance crew on Soyuz T-10. But their lives were made miserable by a shortage of electrical power resulting from problems with the solar arrays.

Soyuz T-10 crew Vladimir Titov and Gennadi Strekalov, fresh from Soyuz T-8 and keen to get started on restoring Salyut 7's electricity, made history on September 27 when their launch vehicle exploded on the pad just 90sec before launch. The launch escape system was used – for the first time in the history of manned spaceflight (it was not used on Soyuz 18-1) – and safely hauled the Soyuz free from the conflagration. The crew landed two miles away, shaken but unhurt.

By mid-October Soyuz T-9 was said to be approaching its 115-day design limit, after which it would be unsafe for a return to Earth, and the Western press abounded in stories of the crew being lost in space. But instead of the keenly anticipated rescue attempt, the Soviets sent up Progress 18 on October 21, putting paid to the rumours for the time being. The cosmonauts made spacewalks lasting a total of 5hr 45min on November 1 and 3 to attach to Salyut 7 the new solar panels ferried up by Progress 18. The result was a substantial boost to the station's power supply. Nonetheless, the crew came home on November 23, "as planned," according to the Soviets.

The STS-8 crew arrive at the Kennedy Space Centre in three of Nasa's fleet of T-38 jet trainers. Left to right: Bluford, Brandenstein, Truly, Gardner and Thornton (Nasa)

Name: STS-8 (OV-99 *Challenger* 3)
Sequence: 106th astro-flight, 93rd spaceflight, 90th Earth orbit
Launch date: August 30, 1983
Launch site: Pad 39A, Kennedy Space Centre, Merritt Island, USA
Launch vehicle: SSME/ET/SRB
Flight type: Satellite deployment in Earth orbit
Flight time: 6 days 1hr 8min 40sec
Spacecraft weight: 242,732lb
Crew: Capt Richard Harrison Truly, 45, USN, commander
Cdr Daniel Charles Brandenstein, 40, USN, pilot
Lt-Cdr Dale Allan Gardner, 34, USN, mission specialist
Lt-Col Guion Stewart Bluford, 40, USAF, mission specialist
William Edgar Thornton, 54, mission specialist

A spectacular night launch into a thunder-filled sky 12 minutes late on August 30 heralded the beginning of a mission that Shuttle programme chief James Abrahamson would later describe as "fabulous". The second Tracking and Data Relay Satellite should have been on board *Challenger*, but upper-stage problems meant that it had to be replaced with tho Payload Flight Test Article (PTFA), a giant dumb-bell designed for use in tests to assess the ability of the RMS to handle heavy payloads. Also in the payload bay was Insat 1A, an Indian communications satellite; this was successfully deployed a day after launch. Mounted in the mid-deck was the Continuous Flow Electrophoresis in Space (CFES) system, making its fourth spaceflight and processing living cells for the first time. There was also a student experiment on

board, plus 260,000 US Postal Service philatelic first-day covers to be sold after the flight. In addition to the five crew, which included the first black, Bluford, and the oldest man in space, Thornton, six rats went along for the ride in a test enclosure. The mission ended at Edwards with the first night landing in the US space programme. Later analysis of the recovered SRBs revealed a potentially fatal fault in one of the nozzles which resulted in delays to STS-9.

Above The first night launch in the Shuttle programme (Nasa)

Below The first night landing in the Shuttle programme (Nasa via Astro Information Service)

The first six-person crew in history was led by the first man to make six spaceflights, John Young (right). The others are, left to right, Robert Parker, Owen Garriott, Byron Lichtenberg, Ulf Merbold (waving) and Brewster Shaw (Nasa)

Name: STS-9 (OV-102 *Columbia* 6)
Sequence: 107th astro-flight, 94th spaceflight, 91st Earth orbit
Launch date: November 28, 1983
Launch site: Pad 39A, Kennedy Space Centre, Merritt Island, USA
Launch vehicle: SSME/ET/SRB
Flight type: Earth orbit
Flight time: 10 days 7hr 47min 23sec
Spacecraft weight: 247,619lb
Crew: John Watts Young Jr, 53, commander
Major Brewster Hopkinson Shaw Jr, 38, USAF, pilot
Owen Kay Garriott, 53, mission specialist
Robert Allan Ridley Parker, 46, mission specialist
Ulf Merbold, 42, payload specialist
Byron K. Lichtenberg, 35, payload specialist

Spacelab was conceived in 1969 when Europe was invited by Nasa to participate in post-Apollo programmes. Development cost to the European Space Agency was £416 million, of which West Germany contributed 54%. Spacelab 1 consisted of a long module and single pallet mounted in the Shuttle payload bay and weighing a total of 33,584lb. Some 22

separate international investigations comprising 72 experiments were carried out in five broad areas of scientific research: life sciences, atmospheric physics, Earth observations, astronomy and solar physics. The crew, commanded by premier astronaut John Young, making a unique sixth spaceflight, included a payload specialist from West Germany. Divided into two three-man teams, the crew worked 12hr shifts for over ten days, one day and five orbits longer than originally planned as a result of a computer failure before retrofire. As *Columbia* came into land at Edwards, leaking hydrazine fuel from an APU caught fire and continued to burn unnoticed after the landing. Only one thing rankled following this very successful mission: under the terms of its contract with the US space agency, ESA had to yield control of Spacelab to Nasa after its first orbital flight.

Above *Spacelab 1 in the payload bay of* Columbia (Nasa)

Below Columbia *comes home after the longest Shuttle mission* (Nasa)

Name: Shuttle 10/STS 41-B (OV-99 *Challenger* 4)
Sequence: 108th astro-flight, 95th spaceflight, 92nd Earth orbit
Launch date: February 3, 1984
Launch site: Pad 39A, Kennedy Space Centre, Merritt Island, USA
Launch vehicle: SSME/ET/SRB
Flight type: Earth orbit, satellite deployment
Flight time: 7 days 23hr 15min 54sec
Spacecraft weight: 250,452lb
Crew: Vance DeVoe Brand, 52, commander
Lt-Cdr Robert Lee Gibson, 37, USN, pilot
Capt Bruce McCandless II, 46, USN, mission specialist
Ronald Erwin McNair, 33, mission specialist
Lt-Col Robert Lee Stewart, 41, US Army, mission specialist

Above The STS 41-B crew pose for the now traditional in-flight photograph. They are, clockwise from top left, Robert Gibson, Vance Brand, Bob Stewart, Bruce McCandless and Ron McNair (Nasa via Astro Information Service)

Below Challenger *becomes the first spacecraft to land back at its launch site, touching down within sight of the immense Vehicle Assembly Building at Kennedy (Nasa)*

Bruce McCandless, the first human satellite (Nasa via Astro Information Service)

STS 41-B, saddled with a new and extremely confusing nomenclature, featured a well deserved flight at last for long-serving Bruce McCandless, a space rookie since 1966. The reason for his long wait was his assignment to development of the Manned Manoeuvring Unit (MMU), which he flew for the first time in orbit during this mission. Flying untethered as far as 300ft away from *Challenger*, McCandless made two MMU sorties, as did his fellow mission specialist, the first Army astronaut, Bob Stewart. With the aid of Ron McNair operating the RMS, both EVA crewmen practised space repair work. The spectacular success of the MMU flights did much to restore US pride following the loss earlier in the mission of two communications satellites. After successful deployment from the Orbiter, the PAM-D upper stages on both Palapa and Westar failed, stranding the satellites in unusable orbits. *Challenger* became the first spacecraft to return to its launch site when it landed on the Shuttle runway at Kennedy after suffering a bird strike on final approach.

Soyuz T-10 February 8, 1984 Flight 109

Name: Soyuz T-10
Sequence: 109th astro-flight, 96th spaceflight, 93rd Earth orbit
Launch date: February 8, 1984
Launch site: Tyuratam, USSR
Launch vehicle: A2 (SL-4)
Flight type: Ferry to Earth-orbital space station
Flight time: 236 days 22hr 49min (landed in Soyuz T-11)
Spacecraft weight: About 15,430lb
Crew: Col Leonid Denisovich Kizim, 42, Soviet Air Force, commander
Vladimir Alexeyevich Solovyov, 37, flight engineer
Oleg Yuryevich Atkov, 34, cosmonaut researcher

If there had been anything seriously wrong with Salyut 7 the station wouldn't have been manned again after the excitements of Soyuz T-9. As it was, Soyuz T-10 was launched on February 8, 1984, with a crew of three cosmonauts who would establish a record stay in Salyut 7 of 235 days, spending some of that time repairing the station. One of the crew was cardiologist Dr Oleg Atkov, flying with a portable ultrasound cardiograph he had designed himself. His job was to monitor the health of the crew during their marathon stay in space.

They received their first visit on February 23, when Progress 19 delivered a cargo which included more medical equipment. (Progress 18 had been de-orbited earlier.) The Soyuz T-11 crew arrived on April 4, and with six people now on board Salyut 7 and five astronauts aboard Space Shuttle *Challenger* on Mission 41-C, a record 11 people were in space at one time. After the guest cosmonaut from India and his two partners had departed in Soyuz T-10, Kizim moved Soyuz T-11 to the front of the Salyut, leaving the rear port free for the arrival of Progress 20 on April 20.

The cargo craft carried equipment to enable the cosmonauts to perform a major repair to the station's main propulsion system, which had suffered a leak the previous September, prompting speculation in the West that the Soyuz T-9 cosmonauts were stranded. The repair was accomplished by Kizim and Solovyov in the course of four remarkable EVAs during which they carried out manual operations comparable in difficulty to the work of the Shuttle astronauts who had repaired Solar Max. Progress 20 departed on May 6 and its successor arrived on May 10 with another set of solar panels which Kizim and Solovyov erected on May 18 during a record-breaking fifth mission EVA. The new arrays boosted Salyut 7's power supply to over 4kW.

Progress 21 left the station and No 22 duly docked on May 30 with more film for the Earth-resources cameras and more fuel, which was pumped on board on July 6. Progress 22's departure was followed by the arrival of the crew of Soyuz T-12 on July 18, two of whom made an EVA from the station on July 25. The Soyuz T-12 crew brought with them an instructional videotape showing Kizim and Solovyov how to carry out further repairs to the Salyut's propulsion system. This they did during a record sixth EVA on August 8.

Progress 23 came and went very quickly between August 16 and 26 and the cosmonauts continued their odyssey, exceeding the Soyuz T-5 duration record of 211 days on September 7. Their journey finally came to an end on October 1, when the frail, pale and tired crew were lifted from Soyuz T-11 and placed in reclining chairs to begin the difficult readaptation to gravity. The cosmonauts had completed 500 experiments and spent over 22hr working in open space. Dr Atkov had spent 87 days conducting medical check-ups.

Above *Kizim (left), Solovyov (centre) and Atkov after their record-breaking spaceflight (Tass)*

Left *Left to right: Solovyov, Atkov and Kizim in a television picture from Salyut 7 (Tass)*

Name: Soyuz T-11
Sequence: 110th astro-flight, 97th spaceflight, 94th Earth orbit
Launch date: April 3, 1984
Launch site: Tyuratam, USSR
Launch vehicle: A2 (SL-4)
Flight type: Ferry to Earth-orbital space station
Flight time: 7 days 21hr 40min (landed in Soyuz T-10)
Spacecraft weight: About 15,430lb
Crew: Col Yuri Vasilyevich Malyshev, 42, Soviet Air Force, commander
Gennadi Mikhailovich Strekalov, 42, flight engineer
Sqn Ldr Rakesh Sharma, 35, Indian Air Force, cosmonaut researcher

Rakesh Sharma became India's first man in space, the third most famous person in India behind Mrs Gandhi and a film star, the 11th Soviet Intercosmos crewman and the 138th man in space a few minutes after his launch in Soyuz T-11 with Yuri Malyshev and Gennadi Strekalov at 5.09 pm Moscow time on April 3, 1984. Docking with Salyut 7 took place 25hr 22min later. For the first time ever there were six people aboard a Salyut, the resident crew of which were spending their 55th day in space. Sharma carried out detailed Earth-resources photography of India and many zero-g adaptation tests, including yoga. The visiting trio departed on April 11, landing at 2.50 pm Moscow time.

Right The Soyuz T-11 crew pictured inside the Soyuz simulator at Star City: Rakesh Sharma (right), Yuri Malyshev (left) and Gennadi Strekalov (background) (Tass)

Name: Shuttle 11/STS 41-C (OV-99 *Challenger* 5)
Sequence: 111th astro-flight, 98th spaceflight, 95th Earth orbit
Launch date: April 6, 1984
Launch site: Pad 39A, Kennedy Space Centre, Merritt Island, USA
Launch vehicle: SSME/ET/SRB
Flight type: Earth orbit, satellite repair
Flight time: 6 days 23hr 40min 5sec
Spacecraft weight: 254,254lb
Crew: Capt Robert Laurel Crippen, 46, USN, commander
Francis Richard "Dick" Scobee, 44, pilot
Terry Jonathan Hart, 37, mission specialist
George Driver Nelson, 33, mission specialist
James Douglas Adrianus van Hoften, 39, mission specialist

The STS 41-C crew pictured at Edwards AFB after the mission. Left to right: van Hoften, Scobee, Hart, Nelson, Crippen (Nasa)

This brilliant demonstration of just what man was capable of in space was made possible by the fact that the defective Solar Maximum Mission satellite, launched in 1980, had been designed for retrieval by the Shuttle. Following the launch, mission commander Crippen came within a second of using up vital OMS fuel for the rendezvous when the SRBs did not perform as well as expected. Then George Nelson's TPAD docking device, with which he was to grapple Solar Max during an EVA with the Manned Manoeuvring Unit, failed to engage properly. The astronaut vainly tried to stabilise the satellite manually to set it up for capture by the RMS, but succeeded only in making it wobble even more. Fortunately,

Terry Hart's precision handling of the RMS the following day resulted in a capture at the first attempt. Solar Max was mounted in the payload bay and repaired during two impressive EVAs by Nelson and James van Hoften. It was then redeployed successfully. The mission also featured the deployment of the largest object yet carried by Shuttle: the Long Duration Exposure Facility (LDEF) carried 57 passive experiments designed for exposure to space and subsequent retrieval and analysis. Crippen, looking forward to another attempt to land at Kennedy, was again thwarted by bad weather and had to settle for a problem-free touchdown at Edwards.

Soyuz T-12   July 17, 1984   Flight 112

Name: Soyuz T-12
Sequence: 112th astro-flight, 99th spaceflight, 96th Earth orbit
Launch date: July 17, 1984
Launch site: Tyuratam, USSR
Launch vehicle: A2 (SL-4)
Flight type: Ferry to Earth-orbital space station
Flight time: 11 days 19hr 14min 36sec
Spacecraft weight: About 15,430lb

American-style walk to the transfer van for the Soyuz T-12 crew. Left to right: Vladimir Dzhanibekov, Svetlana Savitskaya and Igor Volk (Tass)

Crew: Col Vladimir Alexandrovich Dzhanibekov, 42, Soviet Air Force, commander
Svetlana Yevgenyevna Savitskaya, 35, flight engineer
Igor Petrovich Volk, 47, cosmonaut researcher

By mid-1984 the Space Shuttle programme was only a few months away from a mission due to carry the first woman to make two spaceflights and the first female spacewalker. So it came as no surprise to find that the crew of Soyuz T-12, which docked with Salyut 7 on July 18, included Svetlana Savitskaya, making her second spaceflight. She also duly performed an EVA, on July 25. The spacewalk, with four-mission veteran Vladimir Dzhanibekov, was also the first male-female EVA, stealing yet more thunder from the Shuttle. It lasted 3hr 35min, in the course of which both Savitskaya and Dzhanibekov operated a hand-held general-purpose electron-beam tool capable of cutting, welding, soldering and spraying. The success of this demonstration indicated a level of space tool technology yet to be matched by the USA. The Soyuz T-12 crew stayed aboard Salyut 7 for four days longer than the usual one-week visit, returning home on July 28. Little did Dzhanibekov know that he would return to Salyut 7 aboard the very next Soyuz T, tasked with one of the most dangerous and difficult missions in the history of manned spaceflight.

Television picture of the first spacewalk by a woman (Tass)

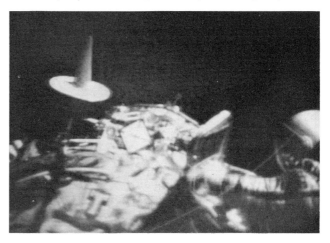

The Discovery *launch pad abort on June 26* (Nasa)

Name: Shuttle 12/STS 41-D (OV-103 *Discovery* 1)
Sequence: 113th astro-flight, 100th spaceflight, 97th Earth orbit
Launch date: August 30, 1984
Launch site: Pad 39A, Kennedy Space Centre, Merritt Island, USA
Launch vehicle: SSME/ET/SRB
Flight type: Earth orbit, satellite deployment
Flight time: 6 days 0hr 56min 4sec
Spacecraft weight: 216,778lb
Crew: Henry Warren "Hank" Hartsfield, 50, commander
Cdr Michael Loyd Coats, 38, USN, pilot
Steven Alan Hawley, 32, mission specialist
Judith Arlene Resnik, 35, mission specialist
Lt-Col Richard Michael Mullane, 38, USAF, mission specialist
Charles David Walker, 36, payload specialist

The 100th manned spaceflight got off the ground only at the third attempt, having suffered the first launch-pad abort in the Shuttle programme. Set for June 25, the launch of Orbiter *Discovery* on its maiden voyage was postponed until the following day as a result of a computer problem. When the launch crew tried again, the Main Engine start-up sequence was halted automatically after 2.6sec when engine No 3 failed to ignite. Engine No 2, which had sprung to life in the meantime, had to be shut down and the whole stack hurriedly made safe. An anxious Nasa cancelled the next flight and combined the work of two flights into one for *Discovery*'s next attempt. The six-person crew finally lifted off on August 30, though not without another day's delay. Sitting in the mid-deck beneath the Shuttle cockpit was the first industry astronaut, Charles Walker of McDonnell Douglas, who was to operate his company's electrophoresis machine.

The SBS-4 comsat was deployed eight hours after launch, followed by Leasat 1, which was propelled frisbee-style from the payload bay. The successful deployment of Telstar 3C on September 1 created another Shuttle milestone – the programme's first triple satellite launch. A 102ft-long solar array, prototype of a structure that could in the future provide Orbiters with additional electrical power, was erected above the payload bay. But in spite of all this solid technical achievement, most of the publicity focused on an 18in block of ice on the outside of the Orbiter, which had to be removed using the RMS lest it break off during re-entry and damage the spacecraft. Following an uneventful touchdown at Edwards Air Force Base, *Discovery*'s debut was hailed as the best Shuttle flight yet.

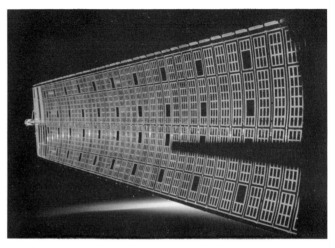

Above *The prototype of the extendable solar sail, seen from one of* Discovery*'s windows* (Nasa via Astro Information Service)

Below *The 41-D in-flight photo is dominated by the legs of America's second woman in space, Judy Resnik. The male supporters are, clockwise from top right, Steven Hawley, Mike Coats, Hank Hartsfield, Mike Mullane and Charlie Walker* (Nasa via Astro Information Service)

Name: Shuttle 13/STS 41-G (OV-99 *Challenger* 6)
Sequence: 114th astro-flight, 101st spaceflight, 98th Earth orbit
Launch date: October 5, 1984
Launch site: Pad 39A, Kennedy Space Centre, Merritt Island, USA
Launch vehicle: SSME/ET/SRB
Flight type: Earth orbit, satellite deployment
Flight time: 8 days 5hr 23min 33sec
Spacecraft weight: 241,780lb
Crew: Capt Robert Laurel Crippen, 47, USN, commander
Cdr Jon Andrew McBride, 41, USN, pilot
Sally Kristen Ride, 33, mission specialist
Kathryn Dwyer Sullivan, 33, mission specialist
David Cornell Leestma, 35, mission specialist
Paul Scully-Power, 40, payload specialist
Cdr Marc Garneau, 35, Canadian Navy, payload specialist

This mission, flown by a record-breaking seven-person crew, featured the deployment of the Earth Radiation Budget Satellite and photography by the Shuttle Imaging Radar and Large Format Camera. So successful was the last that some of its pictures were immediately classified by the US military. The two payload specialists on board were Marc Garneau, Canada's first man in space, and Australian-born Paul Scully-Power, an oceanographer. The mission specialists included Sally Ride, making her second flight, and Kathryn Sullivan, who was to make the first spacewalk by an American woman. The crew was commanded by "Mr Shuttle", Bob Crippen, making his fourth flight.

The deployment of ERBS was delayed when the satellite's solar array would not unfurl, but a nudge from the RMS, operated by Sally Ride, soon solved the problem. The spacewalk saw David Leestma and Kathryn Sullivan carrying out the US space programme's first fluid-transfer tests in preparation for the planned repair of Landsat 4 on a later mission. At the end of the flight Crippen made it to the Cape

at last, following a re-entry that could have been a disaster. Examination of the Orbiter revealed that the adhesive attaching up to 4,000 heatshield tiles to the airframe had been affected by the use of a new waterproofing agent injected into the tiles instead of being sprayed on. This treatment, and the stresses of re-entry, had loosened the tiles to a dangerous degree.

Leestma, left, and Sullivan during their EVA (Nasa via Astro Information Service)

Challenger *swoops like a bird of prey on to the KSC Shuttle runway* (Nasa)

Another record is broken as seven people, including two women and Bob Crippen on his fourth Shuttle flight, set off for space. Left to right: Kathy Sullivan, Paul Scully-Power, Bob Crippen, Jon McBride, Dave Leestma, Marc Garneau and Sally Ride (Nasa)

Name: Shuttle 14/STS-51A (OV-103 *Discovery* 2)
Sequence: 115th astro-flight, 102nd spaceflight, 99th Earth orbit
Launch date: November 8, 1984
Launch site: Pad 39A, Kennedy Space Centre, Merritt Island, USA
Launch vehicle: SSME/ET/SRB
Flight type: Earth orbit, satellite deployment/retrieval
Flight time: 7 days 23hr 45min 54sec
Spacecraft weight: 261,679lb
Crew: Capt Frederick "Rick" Hamilton Hauck, 43, USN, commander
Cdr David Mathieson Walker, 40, USN, pilot
Cdr Dale Allan Gardner, 36, USN, mission specialist
Joseph Percival Allen, 47, mission specialist
Anna Lee Fisher, 35, mission specialist

In-flight photo of the 51-A crew. Clockwise from top left: Gardner, Hauck, Walker, Allen and Fisher (Nasa via Astro Information Service)

Gardner chases Westar during the second satellite capture (Nasa)

Triumphant spacewalkers Gardner, left, and Allen hold up a "For Sale" sign after successfully securing Westar and Palapa inside the payload bay (Nasa)

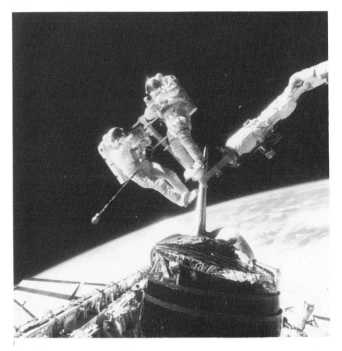

The Palapa and Westar communications satellites, deployed successfully on Shuttle mission 41-B but stranded in useless orbits by faults in their upper stages, gave the 51-A crew a splendid opportunity to show yet again the capability and versatility of the Space Transportation System. The underwriters who had paid out $180 million on the loss of the satellites were paying Nasa a further $10 million for a rescue attempt in the hope of recouping at least some of their loss by re-selling the spacecraft.

Delayed by high winds at altitude on November 7, *Discovery* set off the following day on a busy mission which involved not only the satellite retrieval but also the orbiting of Leasat 2 and Anik D2. These deployments went smoothly and on November 12 the spectacular retrieval began, with Joe Allen, the smallest of the astronauts, flying across to Palapa in an MMU. He docked with the satellite's apogee kick motor nozzle using a "stinger" device mounted on his chest and held the satellite steady while Anna Fisher moved him and Palapa into the payload bay with the RMS. Dale Gardner then began fixing an attachment point to the satellite to permit it to be secured by the RMS while the two astronauts prepared it for mounting in the payload bay. The A-frame attachment proved to be the wrong size, however, so while Allen held the 1,200lb satellite, Gardner fitted the adaptor with which Palapa would be secured in the payload bay. The following day the operation was repeated with Westar. Though

Gardner captured the satellite this time, it was the diminutive Allen who once again held it overhead while his colleague bolted on the adaptor.

51-A, which was launched with two satellites and came home with two more, was hailed as the greatest mission yet and certainly one of the most significant to the future of space exploitation. For their unique salvage work the five Shuttle crew were awarded the Lloyd's Silver Medal.

The first crew of a full military Space Shuttle mission included USAF Manned Space Flight Engineer Gary Payton, standing left. The rest of the crew comprised, left to right, pilot Loren Shriver, commander Ken Mattingly, and mission specialists Jim Buchli and Ellison Onizuka (Nasa)

Name: Shuttle 15/STS 51-C (OV-103 *Discovery* 3)
Sequence: 116th astro-flight, 103rd spaceflight, 100th Earth orbit
Launch date: January 24, 1985
Launch site: Pad 39A, Kennedy Space Centre, Merritt Island, USA
Launch vehicle: SSME/ET/SRB
Flight type: Earth orbit, satellite deployment
Flight time: 3 days 1hr 33min 13sec
Spacecraft weight: 250,891lb
Crew: Capt Thomas Kenneth Mattingly 2nd, 48, USN, commander
Lt-Col Loren James Shriver, 40, USAF, pilot
Lt-Col Ellison Shoji Onizuka, 38, USAF, mission specialist
Lt-Col James Frederick Buchli, 39, USMC, mission specialist
Maj Gary Payton, 36, USAF, payload specialist

The first classified mission of the Space Shuttle programme, STS 51-C received so much pre-publicity that its nature was well known before it left the launch pad in exceptionally cold weather on January 24, 1985. The crew of Orbiter *Discovery* were to deploy a large geostationary electronic monitoring satellite on an IUS upper stage. The military nature of the flight meant that nobody but the launch team knew when the Shuttle would lift off until T−7min, whereupon KSC workers made their customary move to windows and balconies to watch the spectacle.

The five-man crew duly deployed the satellite, which achieved orbit in spite of a performance shortfall by the IUS. This greatly embarrassed Boeing, still red-faced after the TDRS-1 debacle. The mission came to a quiet conclusion after only three days, making it the shortest Shuttle flight so far.

Among the five men who stepped on to the KSC runway after touchdown was the first of a new breed of astronaut who would henceforward travel on all Defence Department missions, USAF Manned Space Flight Engineer Maj Gary

Payton. The possible consequences of launching 51-C in such cold weather were not realised until after the flight, when the recovered SRBs were inspected. What was found should have grounded the Shuttle, but there were to be eight more missions before Nasa finally ran out of luck.

Above *Impressive lift-off for* Discovery (Nasa)

Below Discovery *comes home after the shortest Shuttle mission* (Nasa via Astro Information Service)

STS 51-D   April 12, 1985   Flight 117

Name: Shuttle 16/STS 51-D (OV-103 *Discovery* 4)
Sequence: 117th astro-flight, 104th spaceflight, 101st Earth orbit
Launch date: April 12, 1985
Launch site: Pad 39A, Kennedy Space Centre, Merritt Island, USA
Launch vehicle: SSME/ET/SRB
Flight type: Earth orbit, satellite deployment
Flight time: 6 days 23hr 55min 23sec
Spacecraft weight: 248,927lb
Crew: Col Karol Joseph Bobko, 47, USAF, commander
Cdr Donald Edward Williams, 42, USN, pilot
Stanley David Griggs, 45, mission specialist
Jeffrey Alan Hoffman, 40, mission specialist
Margaret Rhea Seddon, 37, mission specialist
Sen Edwin "Jake" Garn, 52, payload specialist
Charles David Walker, 36, payload specialist

After Mission 41-F had been cancelled in 1984, its crew was transferred to 51-E, scheduled to lift off in February. But then *Challenger*'s cargo for this mission, the second Tracking and Data Relay Satellite, had to be grounded as a result of technical problems and the flight was scrubbed. Finally, Karol Bobko and his crew were assigned to 51-D and given two unique payload specialists: Charles Walker, the first PS to make a second flight, and Senator Jake Garn, head of the Congressional committee responsible for Nasa's budget.

The fourth anniversary of the first Shuttle flight dawned dark and gloomy at the Cape, and it came as something of a surprise to observers to see *Discovery* lift off with just 55sec of the launch window remaining. Anik C1 and Leasat 3 were deployed without incident, but then it became clear that the latter wasn't powering up as it ought. It was thought that the solution lay in tripping an arming pin on the side of the satellite. So, in an unscheduled EVA, the first in the Shuttle programme, astronauts Jeff Hoffman and Dave Griggs fixed jury-rigged "fly-swatter" devices to the end of the RMS. Rhea Seddon then attempted to pull the pin out by snaring it with the fly-swatters while Bobko flew the Orbiter in close formation with the fully fuelled comsat. Although the pin came out partially, Leasat remained dormant and had to be

Above *The crew pose in the white room at the end of the 51-D countdown demonstration test on March 30. Left to right: Jake Garn, Karol Bobko, Don Williams, Charlie Walker, Rhea Seddon, Jeff Hoffman and Dave Griggs* (Nasa)

Below *Hoffman, left, and Griggs carried out an unscheduled EVA to fix "fly-swatters" to the end of the mechanical arm* (Nasa)

abandoned in low orbit. *Discovery* landed in a crosswind at KSC, and Bobko's efforts to keep the Orbiter on the runway centreline by means of differential braking resulted in seized brakes and a burst tyre. Subsequent landings scheduled for the Cape were cancelled, and use of the runway there still had not resumed when the Shuttle programme came to a halt the following year.

STS 51-B   April 29, 1985   Flight 118

Name: Shuttle 17/STS 51-B/Spacelab 3 (OV-99 *Challenger* 7)
Sequence: 118th astro-flight, 105th spaceflight, 102nd Earth orbit
Launch date: April 29, 1985
Launch site: Pad 39A, Kennedy Space Centre, Merritt Island, USA
Launch vehicle: SSME/ET/SRB
Flight type: Earth orbit
Flight time: 7 days 0hr 8min 50sec
Spacecraft weight: 246, 880lb
Crew: Col Robert Franklyn Overmyer, 48, USMC, commander
Col Frederick Drew Gregory, 44, USAF, pilot
Norman Earl Thagard, 41, mission specialist

William Edgar Thornton, 56, mission specialist
Don Leslie Lind, 54, mission specialist
Lodewijk van den Berg, 53, payload specialist
Taylor G. Wang, 44, payload specialist

When scientist-pilot Don Lind was selected as one of the 19 astronauts who joined Nasa in 1966, he had every reason to believe that he would one day walk on the Moon in the Apollo programme. But although many of his contempories did just that, Lind was to be disappointed. In fact he waited a record 19 years before finally reaching space as mission specialist on the Spacelab 3 flight. Illustrating the apparent routine of commuting into space on the Shuttle was the fact that Lind was one of three men over 50 years old on board. Working

with Lind as a mission specialist was the oldest man in space, 56-year-old Dr Bill Thornton, who had to contend with the unfortunate waste management system failures in the mini-zoo installed for a study of spacesickness. Pre-flight misgivings about this "animal holding facility" proved justified when mission commander Overmyer found monkey faeces floating in front of his nose as he sat in his seat in the cockpit.

Work in the Spacelab long module was so highly esoteric that it went over most people's heads, but the massive scientific yield from the flight certainly pleased the experimenters, totalling no fewer than 250,000 million bits of computer data – enough to fill 44,000 200-page books – and three million frames of video. Payload specialist Lodewijk van den Berg operated his own crystal growth experiment and Taylor Wang worked flat-out to repair his malfunctioning fluid mechanics experiment in time to obtain useful results from it later in the flight. The seven-man crew, two monkeys and 24 rats came home to Edwards Air Force Base seven days after lift-off.

STS 51-B crew inside Spacelab 3. Clockwise from top left: Lind, Wang, Thagard, Thornton, van den Berg, Gregory and Overmyer (Nasa)

Soyuz T-13 June 6, 1985 Flight 119

Above *Dzhanibekov, left, and Savinykh arrive at the pad (Tass)*

Left *The launch of Soyuz T-13 (Tass)*

Below *As engineers watch at mission control, Dzhanibekov closes in on the dead space station (Tass)*

Name: Soyuz T-13
Sequence: 119th astro-flight, 106th spaceflight, 103rd Earth orbit
Launch date: June 6, 1985
Launch site: Tyuratam, USSR
Launch vehicle: A2 (SL-4)
Flight type: Ferry to Earth-orbital space station
Flight time:
Spacecraft weight:
Crew: Col Vladimir Alexandrovich Dzhanibekov, 43, Soviet Air Force, commander
Viktor Petrovich Savinykh, 45, flight engineer (landed in T-14)

In early 1985 Salyut 7 was unoccupied, pending another visit from a long-duration crew aboard Soyuz T-13. But in March the Soviets suddenly announced that Salyut had ceased operations. Something had obviously gone badly wrong, and the launch of Salyut 8 was expected to follow. Then, without warning, Soyuz T-13 was launched to Salyut 7 on June 6 with a crew comprising Vladimir Dzhanibekov, on a unique fifth flight for a cosmonaut, and Viktor Savinykh. Their job was to enter the ghost ship and reactivate her. This they did in the course of one of the most remarkable missions in space history.

The two men entered the cold, dead station wearing fur overgarments, gloves, boots and gas masks and over the next five days succeeded in bringing the ship back to life, all the while enduring extremely inhospitable conditions. Electrical circuits were repaired, the solar panels realigned, batteries recharged and life-support systems reactivated. The cosmonauts even carried out a spacewalk on August 2 to work on the panels. Testimony to their great work was the fact that they were still on board by September 18, having received visits from Progress 24 and Cosmos 1669. The following day Dzhanibekov and Savinykh were joined aboard Salyut 7 by the three-man crew of Soyuz T-14. Dzhanibekov returned to Earth seven days later with one of the Soyuz T-14 crew while Savinykh remained on board.

The official crew portrait for 51-G. Standing left to right are Lucid, Nagel, Fabian, Al-Saud and Baudry. Kneeling left is Brandenstein, with Creighton on the right (Nasa via Astro Information Service)

Name: Shuttle 18/STS 51-G (OV-103 *Discovery* 5)
Sequence: 120th astro-flight, 107th spaceflight, 104th Earth orbit
Launch date: June 17, 1985
Launch site: Pad 39A, Kennedy Space Centre, Merritt Island, USA
Launch vehicle: SSME/ET/SRB
Flight type: Earth orbit, satellite deployment
Flight time: 7 days 1hr 38min 58sec
Spacecraft weight: 256,524lb
Crew: Capt Daniel Charles Brandenstein, 42, USN, commander
Cdr John Oliver Creighton, 42, USN, pilot
Col John McCreary Fabian, 43, USAF, mission specialist
Shannon Wells Lucid, 42, mission specialist
Lt-Col Steven Ray Nagel, 38, USAF, mission specialist
Maj Patrick Baudry, 39, French Air Force, payload specialist
Prince Sultan Abdul Aziz Al-Saud, 28, payload specialist

In an imaginative and successful response to increasing commercial pressures from European launcher organisation Arianespace, Nasa began to offer launch customers a seat on the Shuttle for an accompanying payload specialist. Not surprisingly, many countries grabbed the bait eagerly. The first to benefit from this offer was a Saudi Arabian prince, flying to observe the launch of his country's Arabsat 1B. He was Prince Sultan Abdul Aziz Al-Saud, a nephew of King Faud and the youngest person ever to fly on an American space mission. The other payload specialist, France's Patrick Baudry, was flying because Nasa was interested in the space adaptation experiments carried out by Jean-Loup Chrétien aboard Salyut 7 in 1982. Also among the crew was the oldest woman in space, 42-year-old Shannon Lucid.

The mission was described as "easily the best yet". Three satellites – Morelos A, Arabsat and Telstar 3D – were deployed and a fourth, Spartan, was deployed and retrieved after a period of free flight with its X-ray instruments pointed at the Milky Way. 51-G also carried out the first Shuttle experiment related to America's Star Wars programme. Following a farcical start when the Shuttle proved to be turned the wrong way, the crew's work on laser pointing techniques went some way towards showing that powerful lasers could be aimed accurately enough to destroy enemy missiles.

Above *Arabsat goes on its way (Nasa)*

Below *The first mission to carry three nationalities, 51-G included Frenchman Patrick Baudry, left, and Abdul Aziz Al-Saud from Saudi Arabia (Nasa)*

Name: Shuttle 19/STS 51-F/Spacelab 2 (OV-99 *Challenger* 8)
Sequence: 121st astro-flight, 108th spaceflight, 105th Earth orbit
Launch date: July 29, 1985
Launch site: Pad 39A, Kennedy Space Centre, Merritt Island, USA
Launch vehicle: SSME/ET/SRB
Flight type: Earth orbit
Flight time: 7 days 22hr 45min 27sec
Spacecraft weight: 252,855lb
Crew: Col Charles Gordon Fullerton, 48, USAF, commander
Lt-Col Roy Dunbard Bridges Jr, 42, USAF, pilot
Anthony Wayne England, 43, mission specialist
Karl Gordon Henize, 58, mission specialist
Franklin Story Musgrave, 49, mission specialist
Loren W. Acton, 49, payload specialist
John-David Bartoe, 40, payload specialist

Following a launch pad abort on July 12, Mission 51-F got off the ground on July 29, only to suffer the Shuttle programme's first major in-flight emergency 4min 55sec into the launch. One of the Main Engines shut down, requiring the OMS motors to make up the shortfall in thrust and propel

Unusual view of the 51-F lift-off shows the contrast between the almost invisible exhaust and shock diamonds from the three Main Engines and the dense plumes from the Solid Rocket Boosters (Nasa)

Challenger into a low but safe orbit. This procedure – known, perhaps over-dramatically, as an abort to orbit (ATO) – turned out to have been made necessary by a spurious high-temperature warning from an engine that was in fact working perfectly. Moments later mission commander Gordon Fullerton averted a second shutdown, which would have resulted in a potentially catastrophic ditching in the sea, when he recognised another warning for what it was and overrode the system.

Though lower than planned, the orbit was adequate for Spacelab 2's intensive programme of solar physics, astrophysics, plasma physics and astronomical observations. After initially misbehaving, the instrument pointing system, lynchpin of the scientific activities, proved to be extremely successful, yielding what a happy Nasa described as "sensational results". The science crew included the oldest man in space, 58-year-old Karl Henize, who was one of the "Excess 11" astronaut-scientists chosen in 1967.

Above *Fullerton's finger on the button used to command the abort to orbit (ATO). Note the other abort modes – Return To Launch Site, Transatlantic Abort Landing, and Abort Once Around – selectable with the adjoining switch (Nasa)*

Below *Shaven-headed Story Musgrave, centre, is framed by the other Spacelab 2 crew members: clockwise from bottom left, Loren Acton, John-David Bartoe, Gordon Fullerton, Tony England, Karl Henize and Roy Bridges (Nasa)*

51-I's crew framed by the Orbiter entrance hatch. Left to right: James van Hoften, Joe Engle, Mike Lounge, Bill Fisher and Dick Covey (Nasa)

Name: Shuttle 20/STS 51-I (OV-103 *Discovery* 3)
Sequence: 122nd astro-flight, 109th spaceflight, 106th Earth orbit
Launch date: August 27, 1985
Launch site: Pad 39A, Kennedy Space Centre, Merritt Island, USA
Launch vehicle: SSME/ET/SRB
Flight type: Earth orbit, satellite deployment/repair
Flight time: 7 days 2hr 14min 42sec
Spacecraft weight: 237,661lb
Crew: Capt Joseph Henry Engle, 53, USAF, commander
Lt-Col Richard Oswalt Covey, 39, USAF, pilot
William Frederick Fisher, 39, mission specialist
John Michael Lounge, 39, mission specialist
James Douglas Adrianus van Hoften, 41, mission specialist

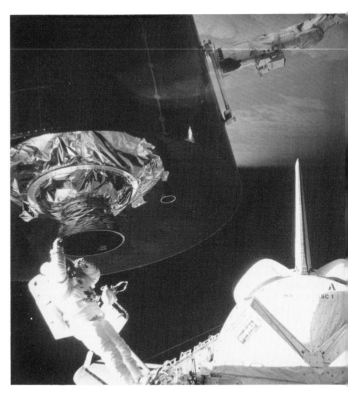

***Above** Fisher installs a protective cover over the engine nozzle of Leasat (Nasa)*

***Below** With a final heave, van Hoften deploys the repaired Leasat 3 (Nasa)*

Before Mission 51-I, it seemed that nothing could cap the Shuttle's earlier achievements in retrieving and repairing the Solar Max, Palapa and Westar satellites. But the mission led by Joe Engle to repair and redeploy Leasat 3 demanded a whole new set of superlatives. At 15,000lb the satellite was huge. The electronics repair was complicated and Leasat was fully fuelled with solid and frozen liquid propellant. The spacewalkers who were to attempt the almost impossible – many people expected failure but nonetheless useful experience – were aptly qualified. James "Ox" van Hoften, the biggest astronaut, would manhandle the satellite, and surgeon Bill Fisher would perform a delicate bypass operation on the satellite's electronics.

The mission began with the deployment of three satellites, Aussat, ASC-1 and Leasat 4. Then, in a two-part EVA on September 1 and 2, the spacewalkers, aided by an RMS operated deftly by Mike Lounge in spite of computer difficulties, captured, repaired and redeployed Leasat with consummate ease. By October the satellite was on station in geostationary orbit and working perfectly. Ironically, Leasat 4, which had earlier reached its orbital slot, suffered a total failure which left it a useless hulk beyond the reach of even the Shuttle's superb repairmen.

Name: Soyuz T-14
Sequence: 123rd astro-flight, 110th spaceflight, 107th Earth orbit
Launch date: September 17, 1985
Launch site: Tyuratam, USSR
Launch vehicle: A2 (SL-4)
Flight type: Ferry to Earth-orbital space station
Flight time: 64 days 21hr 52min
Spacecraft weight: About 15,430lb
Crew: Lt-Col Vladimir Vladimirovich Vasyutin, 33, Soviet Air Force, commander
Georgi Mikhailovich Grechko, 53, flight engineer (landed in T-13)
Lt-Col Alexander Alexandrovich Volkov, researcher

The crew of Soyuz T-14 was an unusual one. At 33, Lt-Col Vladimir Vasyutin was the youngest Soyuz commander for 11 years. Georgi Grechko was brought out of retirement to become the first cosmonaut over 50 to fly, and Lt-Col Alexander Volkov was the first Soviet military cosmonaut researcher since the Salyut 5 military missions in 1977.

After eight days aboard Salyut 7 the crew changed, Grechko returning in Soyuz T-13 with Vladimir Dzhanibekov as the first travellers launched in separate craft to return to Earth together. Resident crewman Viktor Savinykh remained aboard Salyut 7, heading for a space endurance record. This seemed to be within his grasp when on November 15 Soviet news reports on the mission ceased, even though Salyut 7 with its new attachment, Cosmos 1686, which had docked on October 2, appeared to be in no difficulty.

After eight days of stony silence, with the crew apparently communicating in code with ground control for part of the time, Soyuz T-14 was reported to have landed following the first illness of a space crew member to result in an emergency return. Vasyutin had suffered what was described officially as a "psychological trauma". The commander was a "bag of nerves," according to his fellow cosmonauts, who had tried in vain to calm him before being forced to give up and head home. Savinykh had clocked up 168 days in space, which with his Soyuz T-4 experience made him the fourth most travelled space person.

Complications of a crew transfer. **Top** *Soyuz T-14 crew before launch were, left to right: Alexander Volkov, Georgi Grechko and Vladimir Vasyutin.* **Below left** *Soyuz T-13 returned with Vladimir Dzhanibekov, right, and Grechko, while* **below right** *Soyuz T-14 came home with Vasyutin (left), Volkov (right) and Soyuz T-13's Viktor Savinykh (centre)* (Tass)

STS 51-J   October 3, 1985   Flight 124

Name: Shuttle 21/STS 51-J (OV-104 *Atlantis* 1)
Sequence: 124th astro-flight, 111th spaceflight, 108th Earth orbit
Launch date: October 3, 1985
Launch site: Pad 39A, Kennedy Space Centre, Merritt Island, USA
Launch vehicle: SSME/ET/SRB
Flight type: Earth orbit, satellite deployment
Flight time: 4 days 1hr 45min 30sec
Spacecraft weight: About 250,000lb
Crew: Col Karol Joseph Bobko, 48, USAF, commander
Lt-Col Ronald John Grabe, 40, USAF, pilot
Maj David Carl Hilmers, 35, USAF, mission specialist
Lt-Col Robert Lee Stewart, 43, US Army, mission specialist
Maj William Pailes, 33, USAF, payload specialist

The crew of 51-J were: left to right, Stewart, Hilmers, Bobko, Pailes and Grabe (Nasa via Astro Information Service)

Although it was the maiden flight of the new Orbiter *Atlantis*, the military Mission 51-J was by far the quietest, mainly because for once it was truly secret. Virtually all that was known was that the crew would deploy into orbit a pair of DSCS-3 military communications satellites atop a single IUS. The countdown and launch went perfectly and *Atlantis* rose majestically into an almost cloudless sky. The special requirements of the payload dictated an orbital height of 320 miles, the highest ever for a Shuttle. This fact was revealed not by Nasa or the Defence Department but by North American Air Defence Command (Norad). *Atlantis* also carried Nasa's Bios experiment, designed to study the effect of high-energy cosmic rays on biological materials.

   Mission time was just over four days and *Atlantis* came home to a desert runway at Edwards. The five-man crew included USAF Manned Space Flight Engineer William Pailes and commander Karol Bobko, the second astronaut to record three Shuttle missions.

Impressive view of the maiden lift-off of Atlantis *on military mission 51-J (Nasa via Astro Information Service)*

STS 61-A   October 30, 1985   Flight 125

Name: Shuttle 22/STS 61-A/Spacelab D1 (OV-99 *Challenger* 9)
Sequence: 125th astro-flight, 112th spaceflight, 109th Earth orbit
Launch date: October 30, 1985
Launch site: Pad 39A, Kennedy Space Centre, Merritt Island, USA
Launch vehicle: SSME/ET/SRB
Flight type: Earth orbit
Flight time: 7 days 0hr 44min 51sec
Spacecraft weight: 213,070lb
Crew: Henry Warren "Hank" Hartsfield, 51, commander
Lt-Col Steven Ray Nagel, 39, USAF, pilot
Col Guion Stewart Bluford, 41, USAF, mission specialist
Col James Frederick Buchli, 40, USMC, mission specialist
Bonnie Jean Dunbar, 36, mission specialist
Reinhard Furrer, 44, payload specialist
Ernst Willi Messerschmid, 40, payload specialist
Wubbo Ockels, 39, payload specialist

The record STS 61-A crew. Left to right: Wubbo Ockels, Ernst Messerschmid, James Buchli, Bonnie Dunbar, Guion Bluford, Reinhard Furrer (in front of Steve Nagel), and Hank Hartsfield.

A record-breaking eight-person crew took to the skies from Pad 39A on October 30 at the beginning of the first leased mission of the Shuttle programme. West Germany had paid Nasa $175 million to fly its Spacelab D1 and an array of science experiments into orbit. A record number of three payload specialists – two from West Germany and one from Holland, representing ESA – flew on the mission. The commander was Hank Hartsfield, moving up the Shuttle experience league to join Karol Bobko on three missions, and the pilot was Steve Nagel, flying again a record 128 days after his return from 51-G. The trio of mission specialists included one woman, Bonnie Dunbar.

The mission was the first to be controlled from a centre not located in the USA. Transmissions in English, German and Dutch passed between the Shuttle and the control centre at Oberpfaffenhofen in West Germany as the crew worked through an intensive programme of experiments. These included the Space Sled, in which crew members were accelerated up and down a track as their eyes and ears were subjected to various stimuli in an effort to identify the kind of activities and conditions that cause spacesickness. Another novel experiment was a sleeping bag designed by Wubbo Ockels to exert some pressure on the body and so eliminate the sensation of continuous free fall that had made sleeping in weightlessness so difficult for some astronauts.

The highly technical mission, which involved 17 fluid physics, 31 soldification, 17 biological, five medical and six space-time interaction experiments, came to a close at Edwards Air Force Base, where Hartsfield conducted an important nosewheel steering test to clear the way for a planned resumption of landings at Kennedy by Mission 61-C.

STS 61-B   November 27, 1985   Flight 126

Lighthearted 61-B crew portrait shows, clockwise from top left, industry worker Charlie Walker, boss Brewster Shaw, Rodolfo Neri Vela in national costume, barnstorming pilot Bryan O'Connor, and construction workers Woody Spring, Mary Cleave and Jerry Ross (Nasa via Astro Information Service)

Name: Shuttle 23/STS 61-B (OV-104 *Atlantis* 2)
Sequence: 126th astro-flight, 113th spaceflight, 110th Earth orbit
Launch date: November 27, 1985
Launch site: Pad 39A, Kennedy Space Centre, Merritt Island, USA
Launch vehicle: SSME/ET/SRB
Flight type: Earth orbit, satellite deployment
Flight time: 6 days 21hr 4min 50sec
Spacecraft weight: 261,455lb
Crew: Lt-Col Brewster Hopkinson Shaw Jr, 40, USAF, commander
Lt-Col Bryan Daniel O'Connor, 39, USMC, pilot
Mary Louise Cleave, 38, mission specialist
Lt-Col Sherwood Clark Spring, 41, US Army, mission specialist
Major Jerry Lynn Ross, 37, USAF, mission specialist
Rodolfo Neri Vela, 33, payload specialist
Charles David Walker, 37, payload specialist

At the end of what was, yet again, described by Nasa as the "best flight ever," the remarkable Charles Walker, payload specialist for a third time, had clocked up enough Shuttle experience to take him into third place in the league behind Crippen and Hartsfield. At the beginning of the mission

Woody Spring holds the Ease tower while standing at the end of the RMS (Nasa via Astro Information Service)

Walker had more experience of spaceflight than the rest of the crew put together, including flight-experienced commander Brewster Shaw. The second lift-off of *Atlantis*, which was also the second night launch in the Shuttle programme, took place in perfect weather conditions, in contrast with the first night launch, STS-8. The ascending spacecraft could be seen from the Cape until well after SRB separation and by other spectators up to 400 miles away.

Three satellites were deployed flawlessly; they included Satcom Ku-2, on the first uprated PAM-D2 upper stage, and Mexico's Morelos B. On board to watch the Morelos deployment was Mexican payload specialist Rodolfo Neri Vela, a communications expert chosen from among ten finalists.

Because Shuttle flights were now receiving little in the way of publicity, some observers were taken by surprise when Woody Spring and Jerry Ross carried out two of the most spectacular EVAs ever, assembling large structures in the payload bay in the first ever rehearsal of Space Station construction techniques. The Ease/Access experiment was a brilliant success, the astronauts completing their tasks faster and more easily than anticipated, and complaining only of sore fingers. Ross and Spring were the subject of some of the best EVA photographs to come out of the programme. One of these showed Spring at the top of a 45ft-high tower. *Atlantis* came home on December 4 and attention turned again to Pad 39A, whence, two weeks later, the last Shuttle of 1985 was due to take to the skies.

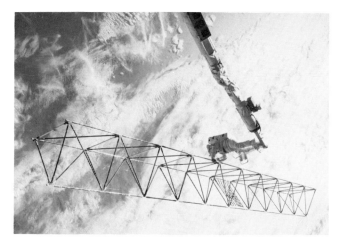

Name: Shuttle 24/STS 61-C (OV-102 *Columbia* 7)
Sequence: 127th astro-flight, 114th spaceflight, 111th Earth orbit
Launch date: January 12, 1986
Launch site: Pad 39A, Kennedy Space Centre, Merritt Island, USA
Launch vehicle: SSME/ET/SRB
Flight type: Earth orbit, satellite deployment
Flight time: 6 days 2hr 4min 9sec
Spacecraft weight: 255,471lb
Crew: Capt Robert Lee "Hoot" Gibson, 39, USN, commander
Lt-Col Charles Frank Bolden Jr, 39, USMC, pilot
Steven Alan Hawley, 34, mission specialist
George Driver Nelson, 35, mission specialist
Franklin Ramon Chang-Diaz, 35, mission specialist
Robert J. Cenker, payload specialist
Congressman C. William "Bill" Nelson, 43, payload specialist

Mission 61-C, the first by Orbiter *Columbia* since STS-9 in November 1983, was routinely postponed for a day from December 18 as a result of a delay in closing out the aft section of the spacecraft. The following day, at T–14sec, the count was stopped dramatically when a back-up hydraulic power unit on the right-hand SRB failed to activate on time. With the Christmas holidays coming up and *Challenger* 51-L already poised for lift-off from the new Pad 39B, Nasa felt able to postpone the 61-C launch to January 4.

But the arrival of the New Year seemed to do little to improve *Columbia*'s chances of getting off the ground. Postponed again to January 6, the launch was scrubbed once more, this time at T–31sec. The following day the crew eased themselves into *Columbia* for what they already knew was going to be a fruitless exercise, so bad was the weather. Two days later the weather was the culprit again. The crew boarded *Columbia* the fourth time on January 9, when at T–9min the weather put paid to the launch once more. By now dubbed "Mission Impossible" by a cynical press, 61-C finally got off to a spectacular pre-dawn start on January 12.

Having experienced the 41-D debacle, mission specialist Steve Hawley now possessed the dubious distinction of having made a total of seven trips to the launch pad for only two lift-offs. The mission itself was the first to carry two crew members with the same surname.

A comparatively spacious schedule called for the deployment of a single satellite, RCA's Satcom Ku-1, and observations of Halley's Comet with a special camera, which failed to work. The flight carried on in this unspectacular manner until it was time to go home. A Kennedy Space Centre landing was needed at all costs to ensure that *Columbia* could be turned around in time for a March 6 lift-off to observe Halley's Comet. Worried about the weather at the Cape, ground controllers first decided to bring *Columbia* home early, then changed their minds twice, and finally ran out of time and had to divert the Orbiter to Edwards for the second night landing of the programme.

The 61-C in-flight photo shows, clockwise from top left, Robert Cenker, Bill Nelson, Charles Bolden, Robert Gibson, George Nelson, Franklin Chang-Diaz and Steve Hawley (Nasa)

Top left Columbia *takes to the skies again after an interval of two years and is still visible* **lower** *as it soars many miles high into the dawn* (Nasa via Astro Information Service)

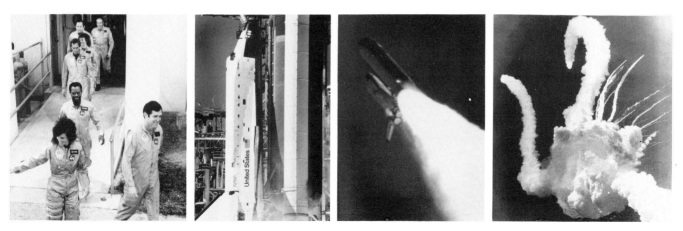

The crew take their final walk. From the rear, Gregory Jarvis, Ellison Onizuka, Christa McAuliffe, Mike Smith, Ron McNair, Judith Resnik and Dick Scobee (Nasa)

Smoke puffs ominously from the joint between the aft centre and aft segments of the right-hand Solid Rocket Booster as Challenger *lifts off on its final flight (Nasa)*

59sec into the flight, a well defined plume of flame can be seen jetting from the failed joint (Nasa via Astro Information Service)

T+73.32sec. The ET disintegrates as its huge load of liquid oxygen and liquid hydrogen is consumed in a near-instantaneous flash fire. The SRBs fly on and will be destroyed 30sec later by the range safety officer. The Orbiter is breaking up as a result of massive aerodynamic overloads (Nasa)

Name: Shuttle 25/STS 51-L (OV-99 *Challenger* 10)
Launch date: January 28, 1986
Launch site: Pad 39B, Kennedy Space Centre, Merritt Island, USA
Launch vehicle: SSME/ET/SRB
Flight time: Vehicle exploded 73sec after launch
Spacecraft weight: About 250,000lb
Crew: Francis Richard "Dick" Scobee, 47, commander
Cdr Michael John Smith, 40, USN, pilot
Judith Arlene Resnik, 40, mission specialist
Ronald Erwin McNair, 35, mission specialist
Col Ellison Shoji Onizuka, 39, USAF, mission specialist
Gregory Bruce Jarvis, 41, payload specialist
Sharon Christa McAuliffe, 37, payload specialist

Originally scheduled to lift off on January 22 and then postponed to January 27, Mission 51-L was delayed yet again by a succession of improbable minor mishaps. A handle fitted temporarily to the entry hatch could not be removed because a bolt had frozen. A drill was called for but its battery was flat. A hacksaw and a second drill arrived, but the bolt was so hard that it blunted the drill bit. Time ran out and the launch was scrubbed.

Nasa could ill-afford this kind of embarrassment, all the more so because the *Challenger* crew included teacher Christa McAuliffe, attracting far more than the usual run of media attention as the first private citizen in space. The following day, January 28, the weather was the coldest ever for a manned launch from KSC. After minor delays *Challenger* made a spectacular lift-off, the first from Pad 39B since ASTP in 1975, and rose into the cloudless sky.

At T+32sec the Main Engines throttled down to 65% for passage through "max Q," the period of maximum aerodynamic pressure, before winding up to 104% at T+52sec. "Go at throttle up" came the call from Houston at T+65sec. "Roger go at throttle up" replied mission commander Dick Scobee. They were to be the last words

from *Challenger*. At T+73sec the Shuttle was blown apart in the most public disaster ever witnessed.

The first US mission to result in in-flight fatalities and the first to leave the launch pad and fail to reach its destination, 51-L made more headlines than any flight since Apollo 11 landed on the Moon. As the investigative process got under way, Nasa's management of the Shuttle was seen to have been "seriously flawed". For example, there had been doubts since 1984 about the integrity of the O-rings sealing the joints between the segments of the SRBs. The space agency had been warned repeatedly to within minutes of the launch that cold weather seriously reduced the O-rings' ability to contain the searing gases generated within the boosters.

But commercial and political pressures prevailed and the 51-L launch went ahead. Videotape replays revealed that a second after lift-off steam and then black smoke began to puff from one of the joints of the right-hand SRB. Thirteen seconds before the explosion a lance of flame from this area played on the External Tank like a blowtorch. The aft link between the SRB and the ET burned through, freeing the big booster to pivot about its top connection, rupturing the giant fuel tank and breaking off part of *Challenger*'s right wing. In the milliseconds that followed, the No 1 Main Engine shut down, probably triggering an alarm in the cockpit. This could have been the last thing Scobee, Mike Smith, Judith Resnik and Ellison Onizuka were aware of before the ET exploded, releasing the boosters to fly on for another 30sec and causing the Orbiter to tumble violently and break up under massive aerodynamic overloads. Below in the mid-deck were Ron McNair, Greg Jarvis and Christa McAuliffe. The crew compartment was later recovered, still containing the remains of the crew.

As the committee responsible for the Presidential investigation into the accident published its findings, confirming an O-ring failure as the cause and recommending design and management changes, the Shuttle was grounded until 1988 at the earliest and the USSR shook down the new Mir space station, setting the stage for a year in space that would surely be dominated by Soviet spectaculars.

Name: Soyuz T-15
Sequence: 128th astro-flight, 115th spaceflight, 112th Earth orbit
Launch date: March 13, 1986
Launch site: Tyuratam, USSR
Launch vehicle: A2 (SL-4)
Flight type: Ferry to Earth-orbital space station
Flight time: 125 days 0hr 1min
Spacecraft weight: About 15,430lb
Crew: Col Leonid Denisovich Kizim, 44, Soviet Air Force, commander
Vladimir Alexeyevich Solovyov, 39, flight engineer

Dressed in their water-cooled underwear, Leonid Kizim, front, and Vladimir Solovyov prepare for an EVA practice in the Hydrolab at Star City (Tass)

A new era in Soviet space exploration and exploitation began on March 13, 1986, with the first Russian manned launch to be shown live on television since the Apollo-Soyuz Test Project in 1975. The Soyuz T-15 launch had been announced the day before and the crew named in a move that was seen as an attempt to capitalise on the Shuttle's public tragedy.

Kizim and Solovyov's target was Mir, a new-generation space station that had been launched on February 20. Mir is a modified Salyut-class vehicle designed to act as the crew quarters of a multi-purpose, permanently manned space complex to be established within four years. It has six docking ports: five at the front, four of which will allow the radial docking of new equipment and experiment modules, and one at the rear. The jubilant cosmonauts opened the doors of Mir on March 15 and showed television viewers around its spacious interior, which had yet to acquire the clutter of equipment typical of a working station. At the same time, the Soviets announced that no more Soyuz T vehicles would fly to Mir, with all subsequent ferry flights to be carried out by a new generation of manned transporters. They also announced that just one more crew would go to Salyut 7 before the earlier station was abandoned.

Progress 25, with two months' supplies, docked with Mir on March 20. The final mission to Salyut 7 turned out to be a visit by Kizim and Solovyov themselves, who performed a historic transfer between two space stations on May 6.

The Mir 1 space station is checked out at Tyuratam (Tass)

On May 21 the Soviets launched the Soyuz TM new-generation manned transporter, which on an unmanned test flight docked with the now vacant Mir 1. Also docked with the station was Progress 26, launched on April 23. On May 28 and 31 Kizim and Solovyov carried out their record seventh and eighth EVAs to perform assembly exercises remarkably similar to the Shuttle Ease/Access operation of the previous November. Soyuz TM returned to Earth on May 28, followed by Progress 26 on June 23. On June 26 the remarkable duo returned to Mir in Soyuz T-15, leaving the Salyut 7-Cosmos 1686 combination vacant again. Then, on July 2, Kizim broke the world space endurance record, four days later becoming the first person to clock up a year's space experience.

It came as some surprise when Kizim and Solovyov returned to Earth on July 16. It appears that delays in the development of work modules for Mir would have reduced the crew to mere sightseeing. Another surprise was the Soviet announcement that Salyut 7 would remain active indefinitely and would receive more crews.

Leonid Kizim pictured on an EVA during the Soyuz T-10/Soyuz T-11/Salyut 7 mission, during which he clocked up 236 days of his world record aggregate of 375 days in space. (Tass)